Disaster Recovery
for LANs

Other Related McGraw-Hill Titles

Disaster Recovery for LANs

A Planning and Action Guide

Regis J. "Bud" Bates

McGraw-Hill, Inc.

New York San Francisco Washington, D.C. Auckland Bogotá
Caracas Lisbon London Madrid Mexico City Milan
Montreal New Delhi San Juan Singapore
Sydney Tokyo Toronto

Library of Congress Cataloging-in-Publication Data

Bates, Regis J.
    Disaster recovery for LANs : a planning and action guide / Regis
J. "Bud" Bates.
        p.   cm.
    Includes index.
    ISBN 0-07-004194-6   —ISBN 0-07-004494-5 (pbk.)
    1. Local area networks (Computer networks)—Security measures–
Planning. 2. Disasters. 3. Emergency management. I. Title.
TK5105.7.B38 1994
004.6'8—dc20   93-23173
                                              CIP

1 2 3 4 5 6 7 8 9 0 DOC/DOC 9 9 8 7 6 5 4 3

ISBN 0-07-004194-6 (hc)
ISBN 0-07-004494-5 (pbk.)

*The sponsoring editor for this book was Jerry Papke, the editing super-
visor was Valerie L. Miller, and the production supervisor was Pamela
A. Pelton. It was set in Century Schoolbook by McGraw-Hill's
Professional Book Group composition unit.*

Printed and bound by R. R. Donnelley & Sons Company.

Contents

Preface   ix

Chapter 1. Introduction   1

Introduction   1
History of Data Processing   2
    The Evolution of Recovery Planning   3
    The PC Emerges   3
    Connectivity Resolved through LANs   4
LANs Defined   4
    What a LAN Is   6
    LAN Boundaries   6
    Functional Parts of a LAN   9
    Equipment Types   10
    Other Components   12
    Servers   14
Topologies   14
    The Star Topology   16
    Bus Topology   16
    Token Passing Ring   19
    Token Passing Bus   20
Bandwidth   22
    Baseband   23
    Broadband   23
Media Used for LANs   24

Chapter 2. The Planning Process   26

Planning for Disaster Recovery   26
    The Need to Plan   30
    What If the LAN Fails?   30
    Planning Strategies   33
Management Commitment   34
    Justifying the Plan to Management   34
The Preliminary Plan   37
The Presentation   38
Options in Developing the Plan   42
    Developing Your Plan   42

Chapter 3. Physical Protection   44

Cable Systems   44
Power Systems   49
Hardware Security   54
Printers   55
Disk Arrays or Disk Servers   57
Other Areas of Concern   57
Software Security   57
Threats   61

Chapter 4. Connectivity Issues   66

LAN to LAN Connectivity   68
LAN to WAN Connectivity   70
Public Network Access   70
Communications Equipment   72
Bridges   74
Routers   79
Gateways   81
Modem Communications   83
   Dial-In Access   85
   Dial-Out Access   91
   Modem Pools   92
   Dial-Back Modems   97
   Leased-Line Modems   101
   Communications Security   104

Chapter 5. Physical Recovery   108

Cable Systems   111
   Cable Cuts   121
   Fire Damage   123
   Water Damage   126
   Rodent Damage   127
   EMI and RFI   130
Media Access Units   131
   Droppage   132
   Electric Spikes   132
   Overheating   133
   Loss of Units   135
Servers   136
   Physical Damage   139
Inaccessibility of Building   139
Terminals, PCs, and Workstations   140

Chapter 6. Departmental Recovery   143

Backup Systems   143
   Disk Mirroring   148
   Disk Duplexing   148
   Tape Backup Systems   150
   Optical Backup Systems   152
   Virtual Disk Systems   155
   RAID (Redundant Arrays of Inexpensive Disks)   160

Network Recovery   161
Communications Recovery   161

Chapter 7. Computer Manufacturer's Involvement with LAN Disaster Recovery   163

Alternative Sites   163
Hotels as an Alternative Site   167
Computer Manufacturers or Hot Site Vendors as Alternative Sites   169
Network Hot Sites   170

Chapter 8. Writing the Plan   174

The Next Steps   174
LAN Standards   178
    Purchasing and Acquisition   178
    Inventories   179
    Virus Prevention   180
    Passwords   180
The Plan   181
Forms   181

Chapter 9. Plan Implementation and Training   198

Schedule the Phases   198
Conduct a Final Review   199
Order Equipment and Facilities   199
Protect the Environment   200
Plan Orchestration and Advertising   200
Training Issues   200
Training Plan   202
Training Materials   203
Testing the Effectiveness of Training   205
The Implementation and Training Plan   205

Chapter 10. Testing and Maintaining the Plan   215

Testing Strategy   215
Who Should Be Involved in Testing the Plan?   217
How Often Should Tests Be Run?   217
Plan Modifications   219
Using Disruptive or Nondisruptive Tests   220
Vary the Tests   221
Plan Maintenance   222
Frequency of Updates   223
Degree of Change   223
Control   224
Distribution   224

Appendix   227
Glossary   237
Index   243

Preface

It never ceases to amaze me that as I travel the country, the risks of disasters on LANs is one of the most significant fears on the minds of LAN managers.

The recent disasters in the United States have exemplified the need for a plan. In a natural disaster (hurricane Andrew) major areas were devastated. The losses were in the billions of dollars, yet many companies were ill-prepared for just such an event. The more recent bombing at the World Trade Center in New York focused on a single complex, yet the losses can easily mount to the billions when all is said and done. Many of the companies that experienced downtime in the mainframe environment had plans in place. Their recovery efforts were easily enacted and the systems were up and running at hot sites or alternative sites.

However, LAN managers were the first to admit that they either did not have a recovery plan or the plan was far too simple to react to a disaster of this sort. Others admitted that if they had a mainframe the process would have been much easier than trying to recover a LAN.

The general feeling is still one that concerns many of us in the industry. First, too many organizations still do not perceive the need for a disaster recovery plan for LANs. This is because the PC-based LAN is viewed as easy to install, throwaway technology. Second, the mission-critical services being migrated onto LANs are overlooked as "necessary" recovery priorities.

The industry has come to the forefront since these two (and other) major disasters. Everyone sees a new "awakening" in disaster recovery planning. Hot site vendors are going to offer new services that address office recovery, specifically LANs. Consultants are offering advice and guidance on how to deal with water- and smoke-damaged PCs, disks, and other media. All noble acts, but too little, too late.

We all must finally wake up to the basic fact that LANs are going to fail and the results can be disastrous. Management must be made aware that this is the life blood of the organization's processing, and funding a plan is mandatory. Ongoing support is no longer optional. Only through this awareness can significant changes be enacted. If these conditions are not met, the renewed interest in disaster recovery planning will fade into the back of people's minds. All effort will cease, since disasters only happen to others.

Until the next one occurs, the cycle will repeat itself.

Acknowledgments

This book is the direct result of the efforts of others who were willing to share ideas, concepts, and war stories. Special thanks go out to the people at Vortex Systems, WBS Associates, and Maynard. Their willingness to help and offer assistance kept the desire to complete this work alive.

Further accolades are required for the people who put the work into providing the text and graphics. Special thanks are in order to Gabriele (Peschkes), the artist who was able to make sense of my drawings and give them life and credence; to Mary Farabaugh, who had to decipher my meanings continually as I rambled on in text, and make everything logical; and to Janet Schuler, who had to jump in at the end and help finalize things. This team put order where chaos reigned; their work is much appreciated.

Regis J. "Bud" Bates

1

Introduction

Introduction

Local area networks (LANs) are an amazing phenomenon in business today. More LANs are being installed on a daily basis than any other computing network architecture. The reasons for this deployment are many; however, some of the primary considerations for installing a network are as follows:

1. Users want *more* control over their data.
2. Off-the-shelf products are readily available.
3. LAN operating systems (OS) are maturing at an accelerating rate, offering the user more control.
4. Mainframe computing resources are expensive, despite the dramatic decrease in costs of computing and storage. The PC (personal computer)-based network offers the equivalent processing power of a mainframe at what appears to be lower costs.
5. Pent-up frustrations with large management information systems (MIS) shops that cannot deliver the applications to the user in the time frames expected.
6. The highly visible LAN marketplace offers to solve the user's demands and problems in a quick time frame.
7. Lack of understanding by departments, management, and vendors alike exists. In each case a substitute network to the large mainframe or midrange computing facilities offers solutions to immediate problems. Since vendors sell a complete package (the PC hardware, network interfaces, network operating systems, and PC applications), they deal with novices in the LAN arena. This leads to a "bill of goods" approach that can leave the end user in a precarious situation. Vendor selection and support, although critical, are often overlooked by the inexperienced user.
8. Outsourcing, downsizing, or lack of management budgetary support for the mainframe compels the MIS department or individual user department to scurry around for a suitable replacement.

9. Workgroups in the organization that share tasks and resources to accomplish their mission are on the rise. Most of these are dynamically formed, requiring flexible and transparent connectivity.

There are other driving forces too numerous to list. The gist is that LANs are being installed as the panacea for the organization's computing needs. Although this may be true, the final acceptance of LAN technology and applications still rests with the ability to perform over the long term and continue to meet the organization's needs. These needs encompass

1. Highly reliable hardware platforms.

2. Faster and more efficient processing.

3. Sufficient throughput at the interface and physical cable to reduce or prevent bottlenecks.

4. Clean operating systems software to allow the user maximum flexibility.

5. Cost containment capabilities that reduce the overall processing expenses.

6. Highly secure operating environments that minimize risk of external threats such as viruses, hacker penetration, destruction of data, and loss of information. Furthermore, internal security to limit access to critical or confidential information is required as more applications are migrated onto the LAN.

As a means of defining and meeting the objectives stated above, the evolution and historical concepts of processing information should be recapped.

History of Data Processing

Before putting a lot of effort into preparing a disaster recovery plan for a LAN, we should understand the historical perspective of what LANs *are* and *why* they evolve in an organization. In the 1960s the processing of data within an organization was handled by the data processing department and is now renamed management information systems in many organizations. You remember the old saying about the "glass house" (you could look in and see but you couldn't touch)!

In the 1960s the data processing departments realized the critical aspects of protecting the organization's data and the need for security, standards, policies, etc. Since they were chartered with both the processing and the protection of information, they rose to the pinnacle of protectionism. No one entered their environment without a true need to gain access. Furthermore, inexperienced users and the lack of computer training exposed systems hardware and critical data. Therefore, only qualified, trained data processing personnel were allowed to input the company's data into the computer. Although this frustrated users, the data processing staff stood firm in their convictions.

Proof of their abilities to orchestrate the processing and protection of company information came as the separation of machine from humans was instituted. The data processing departments built their environment in a secure room, controlled the input/output, and set priorities for the business schedules. They took control seriously! Additionally they provided all of the recovery processes by creating backup files and attended to maintenance and diagnostics, and prevented unauthorized access to rooms, files, applications, etc. They did it all! No one touched the equipment or systems without their approval; hence, the glass house.

The evolution of recovery planning

The data processing departments recognized the need for a recovery plan for the company's data and the major investment made in these mainframe computers. They recognized that all processing within a company could cease if a single failure occurred. Thus they formed specialty teams with expertise in the hardware, software, and communications aspects of data processing recovery. They created a formalized structure as follows:

1. What to do
2. When to perform a task
3. Who should do it

Then they tested the recovery process on a regular basis, documenting it fully. They fought for and won the support of management to plan for the worst.

In the 1970s the situation began to change. Computing was leading the trend to information management. The environment also changed! Departments within an organization demanded and were given the means to input/output their data and to control their own files. The terminals that had previously been housed in a secure environment found their way to specific desktops. Users were becoming more literate and demanded more services from data processing. The walls began to shake in the "glass house" as the end users began to take limited control of their data away from data processing.

Soon after the terminals were placed on the desktop, newer communications capabilities began to pressure the data processing staff to go beyond the initial boundaries of the department or building and penetrate remote sites! Users based 5 or 3000 miles away could control their data.

A new revolution had begun. Users were learning about computers and demanding new applications, faster service, shared access to files, and more control over their information. The walls began to tumble down in the "glass house." The computer was rapidly becoming a tool to increase productivity in the office, which created heavier demands on data processing staffs to be more flexible in dealing with their customers.

The PC emerges

The 1980s brought the introduction of a lower-cost computing service to the desktop. IBM introduced the personal computer, which allowed the user to choose whether to access the mainframe or to process information locally. The umbilical cord to the mainframe as the only device that housed the user data began to weaken. The PC became an island of computing power, isolating users from files and applications on the host system but accessible to specific applications that were rapidly developing for the end user. Applications such as word processing, spreadsheets, and databases allowed individuals to create and control information and manipulate their own data in ways that had never been available to them before. The frustrations over the lack of cooperation and the time it took to get things done within data processing led more users to pursue alternatives. Although the PC was expensive in the early days, the end user was willing to pay the price to regain control.

However, the users quickly learned that they still needed to connect to the host system, so connectivity became an issue in the late 1980s. Users wanted control over their data and the applications, but needed access to other company records and files to perform their normal functions. The need to share files, applications, and expensive resources (printers, etc.) also became an issue. Although these issues surfaced before the 1980s, deployment and acceptance boomed especially during the 1980s.

Connectivity resolved through LANs

The industry responded with a connectivity solution that allowed for local computing as well as file sharing. Called local area networks, a new wave began. Once connectivity and sharing could easily be resolved, the need to have data processing control the user input/output, applications or resources diminished rapidly. As a matter of fact, the data processing staff was viewed as the bottleneck! Everything a department wanted was either negatively received by data processing or was placed on a two year or greater delivery cycle. Neither of these two approaches satisfied the rising demand. Users felt compelled to find off-the-shelf products that could meet their information processing needs while allowing ease of use. Downsizing of the mainframe staff followed quickly and the dependency on the data processing department began to wane.

This happened so quickly that the data processing staffs could not control or manage the influx of PCs, applications, or connectivity requests. Users therefore began to buy and install their own systems. Most of these systems were bought by inexperienced managers who refused to check with MIS or data processing staff.

Unfortunately many department managers bought into technologies or applications that were incompatible with the rest of the company. The LAN had arrived but brought with it a series of related problems. Furthermore, the user and owner of a LAN did not understand and therefore did not take precautions against the loss of data or loss of processing capabilities. They were not trained to diligently back up the disk and critical files as the data processing staff had been drilled originally. The "fun" was about to begin.

LANs Defined

In order to understand the needs of a LAN owner or manager a quick review of the pieces involved and what constitutes a LAN is in order. Defining a

LAN is always a challenge. Ask a dozen users, and you get a dozen different answers. So let us make the following attempt.

A LAN is a phenomenon that has evolved from a centralized data processing facility. It is a shared transport of information arrangement at either the workgroup, department, or company level. Sharing begins with the medium (or cable system):

1. A single cable is run from device to device providing a common path for all equipment and users. This provides the physical transmission capability.

2. Resources are shared such as disk drives for storage; files for access to information; applications that multiple users need to run; and output services, such as printers, modems, and plotters.

3. A contention-controlled service allows a single user on a cable at one time. Another form of LAN supports multiple users on a single cable that is partitioned using either time- or frequency-division multiplexing techniques. In order to provide the smooth and orderly use of these shared resources, an access control scheme allocates the use of resources, thereby preventing corruption of data due to any conflicts of transmissions. This process operates at the data link level using a media access control (MAC) and a logical link control (LLC) function to provide the resolution of contention problems that might arise.

4. Since the LAN uses a shared facility (the cable) and resources, control of the network is not maintained by a single device. Moving away from the centralized control of a mainframe, LANs pass control from device to device. Theoretically no single point of failure should exist that would either limit or stop the use of the LAN. However, in reality single points of failure reside in several places on a LAN. This is the crux of many of the problems that will occur on the LAN.

Table 1.1 is a summary of the features of a LAN as defined above. It behooves planners and administrators to keep these factors in mind when laying out any

TABLE 1.1   Summary of LAN Features

Definition	Purpose or intent
Shared communications facility	A single cable used by all
Shared resources	Printers, plotters, lasers Modems, other communications access Hard disks, direct access storage devices (DASD) Files by user, workgroup, department Applications programs
Contention control	Requires the right to use to prevent corruption of data during transmission
Local area	Geographically bounded by some limitations
No common control	Access is available to all users Control of network is passed from device to device via access control system No single point of failure

new LAN or rearranging an existing LAN. Keeping these features in the forefront may well change the design and layout function by these personnel. Unfortunately when these criteria are ignored, the manager or administrator of the network becomes subject to constant failures and frustrations. Moreover, the layout that is designed and installed incorrectly poses far more difficulties when changes are necessary. These difficulties are represented in the loss of employee confidence in the networks' capabilities, loss of management commitment and support, added costs to change or rearrange the network, and the inability to predict performance statistics accurately for fine-tuning purposes.

What a LAN is

1. A shared connectivity scheme or a single cable shared by multiple users.

2. Locally controlled shared resources—the department controls the input of information and shares resources such as

   - Disks
   - Files
   - Applications
   - Printers, plotters
   - Communications

3. Local area—usually a department, floor, or building, but for this discussion it will be bounded to a mile of cable.

4. No common control—no one station or device controls access or service on the LAN.

LAN boundaries

By sheer definition LAN is a local area service. But just what does local area mean?

To explain this in a usable sense, every organization's interpretation would be different, based on the layout of their facilities. Local can mean a small workgroup clustered together in an open office environment. Figure 1.1 is a representation of such a network. In this particular scenario, all users in a workgroup (payroll, for example) in the accounting department, are linked together. This allows payroll employees access to files and employee pay history while keeping the information under "local" control. Any member of this LAN has access rights to the files and payroll records but no access is granted to anyone outside the boundaries of this group.

Should access to information on this system be required by others, assuming they have a need to know, they would have to enter the workgroup to access the files from one of the workstations (or PCs) attached to the LAN.

Regardless of how these users are attached, whether on a ring or a bus (see the discussion in the section titled Topologies) the idea is the same. In order

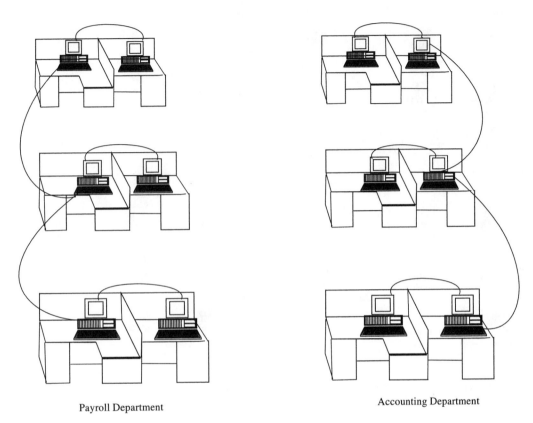

Payroll Department

Accounting Department

Figure 1.1 A workgroup within a department (i.e., payroll) connected together on a LAN.

to access information the user must be on one of the attached devices. Thus we can reasonably assume a local environment in this particular group.

Another possibility is a whole floor of a high-rise office building, which could be multiple departments or the entire company, connected together via a LAN. Figure 1.2 is a representation of this scenario. Note that in this figure, a few changes have occurred. Specifically, multiple departments are linked together, all of the cables are "home run" to a centralized closet where the connectivity actually takes place, and multiple servers exist at various points on the network. This shows the benefits of linking the entire floor onto a single shared system. As cross-departmental needs arise, the user with access privilege can use any service on the network.

Once again, the connection scheme here is for representative purpose only. Each organization may connect these users together using different techniques.

Clearly local area is defined differently in both of these scenarios. For purposes of defining "local" in this book, we will attempt to use a bounded area of less than 1 mile. This 1 mile distance is a cable limitation of some of the

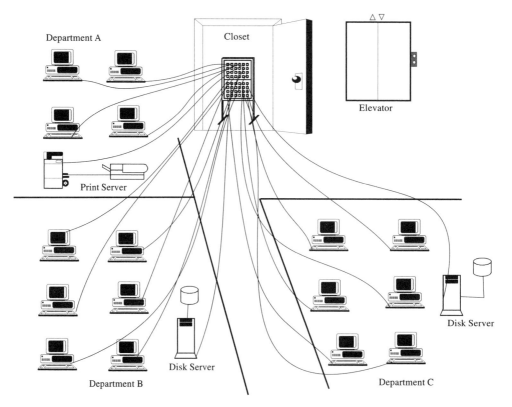

Figure 1.2   A whole floor in an office building on a LAN connecting multiple departments.

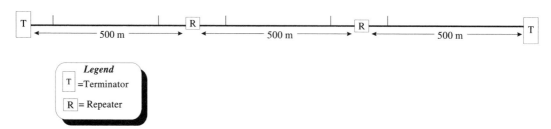

Figure 1.3   Extending cables with repeaters to approximately 1 mile.

networks that exist on the market. An example of this would be to use the Ethernet* definition based on a coaxial system. The Ethernet standard states that each cable segment is limited to 500 m (see Fig. 1.3). When using repeaters (to extend the cable distance and boost the signal strength) a mes-

---

*Ethernet is a registered trademark of DEC, Xerox, and Intel.

sage shall pass through no more than two repeaters. With this limitation taken into account a network could have three segments connected via two repeaters, which extends the actual backbone to a maximum of 1500 m. This is close enough to a 1 mile distance for the examples used. Many of these distance limitations are used to ensure the necessary timing to propagate the signal to the ends of the cable. Thus, the one mile distance can be safely used when trying to define the boundaries of the LAN.

One quick note here is to remember that the actual cable length is the limitation. A straight cable laid out 1 mile long is what the bounding is all about. When installing a cable on multiple floors in a high-rise office building or in a low-rise factory the cable may be snaked, twisted, and turned in many directions. Consequently, depending on the way the cable is run, the distances between the two terminated ends of the cable may appear much closer, even though the actual cable is 1 mile long.

Functional parts of a LAN

Traditionally the functional parts of a LAN include the reason for installing the network in the first place. The original purpose was to accommodate the end user's need to share information within a "community of interest" grouping. This evolution came as the result of the pent-up demand and frustration with data processing departments, as already mentioned. However, users were already using techniques to get around the lack of connectivity, commonly known as the following.

1. "Sneaker-net": the walking of information, files, or floppy disks down the hall to another user

2. "Floppy-swappy": the trading of information strictly on diskettes, and/or using a file transfer from one PC to another via modems and modem eliminators

Users were merely trying to share their files or work product with other interested parties. The stand-alone PC did not allow for real-time sharing. As we all know, these methods above were only partially effective. Once any user touched the data, everyone else's was out of date from that moment on. Several versions of the same data existed.

The user wanted to share information typically involved with the following.

Documents	Very lengthy product specifications, narrative files, and other large items that were manipulated by many interested parties who shared joint authorship or responsibility.
Memos	Short administrative reports and memoranda that update users on the status of projects, meetings, etc. These were supplemented with electronic mail messages as items of "point of need" informational updates.
Scheduling and information	Project plans that multiple parties had input into; calendaring information and appointment setting.

Accounting	Spreadsheets on projects, business memoranda, and actual accounting data (i.e., payables, receivables).
Engineering diagrams	Computer-aided design or manufacturing (CAD or CAM) type files that must be reviewed by various trades of professional designers. This technique allowed for electronic updates to drawings, thereby supplementing blue line drawing and reproduction. Imagine how this technique helps a major construction effort where multiple revisions to drawings, floor plans, electrical pulls, plumbing runs, etc., are constantly in a state of flux as the dynamics of construction are ongoing.

To support this information transfer among users, many specialty devices were purchased. These devices were placed on the desk (or in a department) of specially trained users who controlled the input and output of the information. When a shared device was installed due to financial or space constraints, then queuing began to occur. More users wanted access than devices were available. This was a recurrence of the early data processing days. To overcome the queuing problem managers began to support LAN technology. Unfortunately, many of these LANs were incompatible with others already installed. To work around this, they began to open the systems up to everyone by allowing anyone to access the LAN with their own device or by allowing access to the data through modem access from outside devices.

Equipment types

As LANs were deployed the types of equipment to be installed became another hurdle to deal with. Many organizations already had significant computer systems. With these larger systems the depreciation cycles had not been fully used. As a result, equipment on the books could not be discarded. Consequently, LAN managers had to find ways of attaching the older installed base of terminals, computers, printers, etc. This led to complexities in LAN layouts and difficulties in finding suitable platforms on which to run all the equipment types involved. The typical equipment is comprised of the following:

1. *Computers.* Mainframe, midrange, mini-, and microprocessors are all candidates to act as nodes on a network. Since the LAN emerged from the microcomputer world (PC) the other computers are often used as specialty servers.

2. *Terminals.* Dumb terminals that were traditionally linked to a host system migrated onto LANs. These devices, although they really violate the control factor since they require a common or centralized controller, are often used on LANs when piracy, bootlegging, and threats of virus appear. A user with a dumb terminal cannot load any software onto the network without the assistance of another user who has input capabilities through a floppy disk device. Functionally these devices will be limited to the applications they can

access but for casual users of word processing, database, electronic mail (e-mail), and spreadsheets, they may well serve the need.

3. *Disks.*   Normally the use of disks includes the local hard drive in a PC that can be accessed from the network. Other disks include the larger disks (600 Mbyte, 1.2 Gbyte, and up) which are on a disk server. Furthermore, the direct access storage devices (DASD) or a host computer can be added for storage on the network through special arrangements.

4. *Servers.*   Whether they are stand-alone or shared as a workstation the servers provide a utility function to network users. These devices act as the traffic cops to all the shared resources on the network. They can serve single or multiple functions. Normally they include:

- Print servers
- Disk
- File
- Application
- Communications
- Terminal

5. *Printers.*   Any output device is normally an occasional use system. Users were faced with using inexpensive dot matrix printers for draft quality and the more expensive laser printers for letter quality, final output. To maximize the use of the output device, many users shared a single printer. However, the drivers and setup differences between the draft and letter quality devices caused problems. Size, pagination, and formats were handled differently. So the actual output of these two disparate devices may be completely different. Other devices such as plotters, color printers, slide makers, etc., also fall into this category: too expensive to give to everyone so they must be shared.

6. *Modems.*   Communications from a PC are pretty straightforward today. Scripts and log-ons are standardized to a great extent. Giving every user a modem is similar to the printer problem. The device is usually on an occasional use basis, so it should be shared. Modem pools or communication servers help to bring this sharing about. The modem is now a network resource rather than under an individual's control. Modems are usually set up as inbound or outbound, variations include call back and dial backup systems for leased line modems.

As each of the elements above-noted were discussed, various situations showing advantages and disadvantages were introduced. Detailed discussions of the risks and recovery procedures will be introduced in Chaps. 3 to 5. The devices include a medium or cable system composed of the following:

1. Twisted pairs of wires (shielded or unshielded)
2. Connector cables

3. Fiber optic cables

4. Coaxial cables (thick or thin)

- Baseband
- Broadband

Other components

Still other components are needed to create a LAN. Thus far the pieces included the functional and the primarily hardware pieces. A network consists of a series of interconnecting devices, sharing various resources on a common platform. To get to this common platform the following components are also needed:

1. *A medium.*   The cable system that everyone is sharing to transport the information. Although various forms of media exist, the limitations of each set the standards of how the network will be installed and how many devices will be attached.

2. *A connector.*   A physical link from the device to the cable system. This could be as simple as an RJ-45 telephone jack as shown in Fig. 1.4, or a specialty connector as shown in Fig. 1.5. Again, the means are not as important as compatibility to the overall network.

3. *A network interface.*   Typically a printed circuit board or card installed in an expansion slot in a PC. The card, called a network interface card (NIC), serves multiple functions and is composed of various options:

- A central processing unit (CPU) chip, with its own processor that operates at the network speed. This is the local controller for the delivery and receipt of packets of information within the network.

- A buffer to provide temporary storage while the information is being formatted to or from the network. The buffer varies by card type.

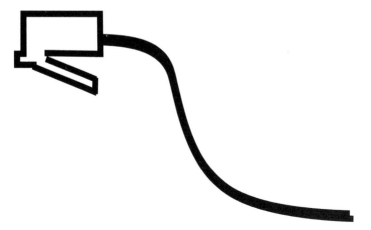

Figure 1.4   RJ-45 plug.

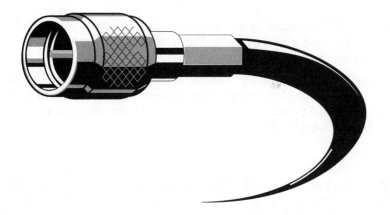

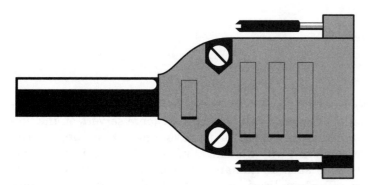

Figure 1.5   Specialty connector plugs.

4. *Media access unit.*   Allows the devices to connect to the backbone network. The media access unit can also be called a multistation access unit (MAU), allowing multiple stations (nodes) to attach to the backbone through a single point. These MAUs usually support increments of eight nodes (8, 16, etc.).

5. *Software.*   The network operating system (NOS) sets the stage for how the network will perform. Features, functions, protocols, etc., are all parts of the NOS.

6. *Options.*   When various networks are installed other devices may be used to provide ease of installation, moves, and changes, and diagnostics. Some of these options may include:

- Patch panels for access to and from the station drop (twisted pair, coax, etc.) to the media access unit. These patch panels allow for ease of installation, moves, and changes.

- Intelligent hubs (newer hubs either active or passive) provide for diagnostics and control. A network manager can electronically provide switching of ports and diagnose problems on the network without traveling to a closet or panel.

Servers

The term *servers* has been used already several times. These devices are typically PCs that serve a utilitarian function. Since the servers on the individual network will vary, the server may serve a single function or multiple functions. Normally the servers are classified by the function they provide. Typical names for these devices include:

- File servers

- Disk servers

- Application servers

- Print servers

- Communication servers

Since many LANs have migrated from a mainframe or minicomputer environment, these older devices may well become very large servers on the network. A typical example of this might be to have a host as an application, disk, and file server combined within a single system. The applications may be "back office" functions such as human resources, payroll, or accounting applications. "Front office" applications might include graphics programs, word processing, spreadsheets, and local databases. These may well reside on separate servers, such as departmental PCs.

Unfortunately, some smaller LANs may also use the server as a workstation. Although this might be more cost effective, since only one PC is used to handle multiple functions, the risks associated with this approach may be more costly in downtime. Think of a station user who gets trapped in an application on the network. This user is on a workstation that also functions as an application server. What happens to the network when this user hits the old reliable "control–alternate–delete" keys to get out of an application? What happens to the open files?

Topologies

A discussion of topologies is introduced here to ensure that the reader fully understands what is going on within the network boundaries.

Each LAN is designed around a structured layout of the cabling systems. How the cables are run sets the stage for the topology. This design takes into account how devices on the network will be interconnected, the intermediate pieces that will be placed on the network, and the flow of the information from device to device.

Older hierarchical systems used a topology that linked devices through a formalized hierarchy. This was the mode of the mainframe computing networks as represented in Fig. 1.6. In the days of mainframe distribution, this hierarchy served its purpose. However, several bottlenecks and points of failure existed in this environment. Fine-tuning was necessary to prevent bottlenecks, increase performance, and improve response times. Consequently the original design of LANs was to move away from this topology. Newer forms of connection were used to meet the needs of the end user. These topologies include the following:

- Star
- Bus
- Token passing ring
- Token passing bus
- Hybrids

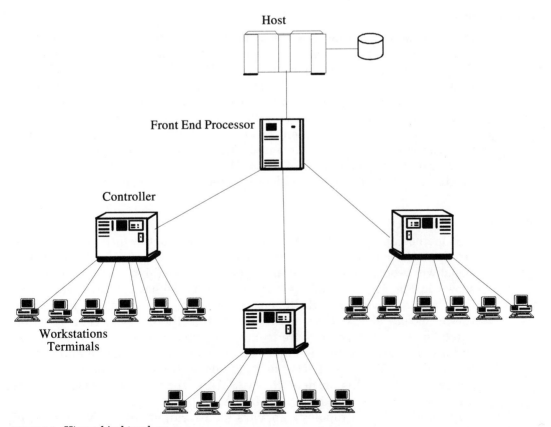

Figure 1.6 Hierarchical topology.

Standards have been developed for the use of these systems by the Institute for Electronic and Electrical Engineers (IEEE) and the International Standards Organization (ISO). These standards provide for the two ingredients of connectivity and topologies:

1. Logical
2. Physical

The difference between physical and logical connectivity will become apparent as the installation and use of a LAN progress. However, simple definitions follow:

- The physical structure is how the cabling system is installed in the building. Physically the cables attach devices such as PCs or workstations to a backbone network for transmission.

- Logically the devices transmit to each other via a broadcast or a passing of information from one to the other.

The star topology

In a star network all devices are "homed" to a control device. Essentially a series of point-to-point connections are installed from a central control point. Connectivity between two devices (A and B) is shown in Fig. 1.7. When A transmits information to B, a central switching arrangement takes place. Since all wires run to this central switching device the connections are fairly straightforward. However, when reflecting back to the definition of a LAN this central switch is a common control point and the single point of failure. Star configurations usually are wired back to a wiring closet (such as a telephone closet) on a floor-by-floor basis. Many times users will attempt to use this topology to be consistent with other existing cable plants in the building.

Bus topology

This method appears to be a straight piece of cable onto which all devices are attached. No central switching or relaying will take place since the electrical signal will propagate along this single piece of cable (highway). In the bus, as shown in Fig. 1.8, the cable is run throughout the environment while taps into the bus provide the attachment of devices as appropriate. The example uses a coaxial cable, but it is not limited to this, therefore the cable is terminated on both ends. Since there is only one transmission path, a contention and access control system is implemented to ensure the smooth and orderly flow of communications. This is normally done through carrier sense multiple access with collision detection (CSMA/CD) or carrier sense multiple access with collision avoidance (CSMA/CA).

Although the bus topology appears as a straight line, other means of installation can be used. Snaking of cable through an office environment can also work as shown in Fig. 1.9. In this example the bus could be snaked

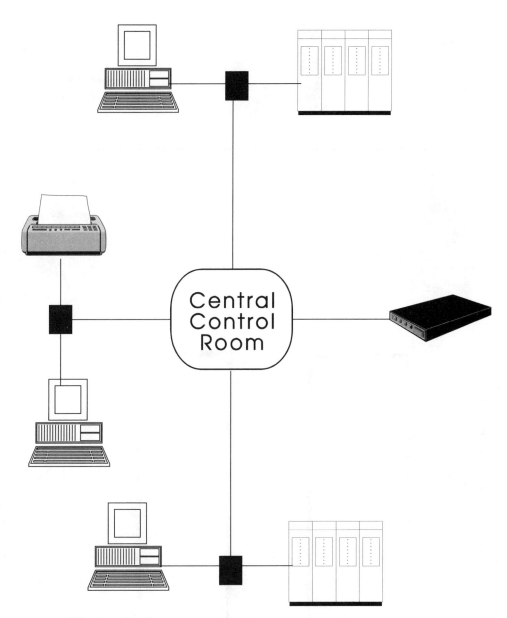

Figure 1.7 The star network.

through a floor tap system or run in the ceiling space. Workstation or nodes would tap into the bus as appropriate.

A third way is to use a physical bus but a logical star (hybrid). In this case, as shown in Fig. 1.10, the bus is run from floor to floor in a horseshoe (inverted U). This is the physical-logical bus, but all taps into the bus are in closets. As the workstations are attached to the bus all connects are "homed" into the clos-

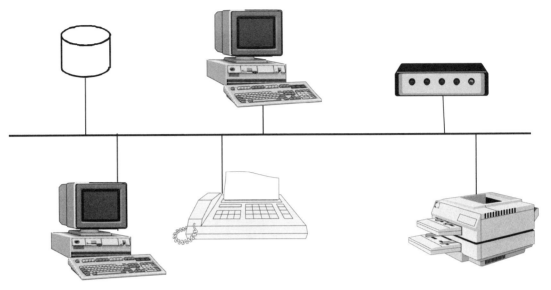

Figure 1.8 The bus topology.

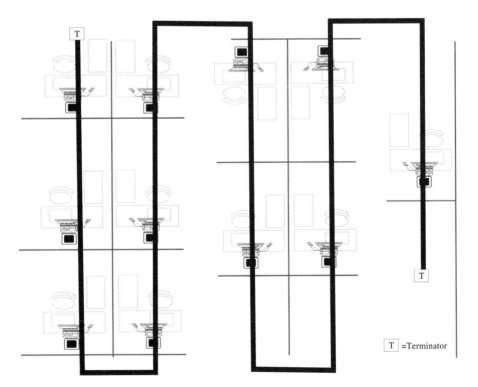

Figure 1.9 The bus cable can be snaked throughout the building.

Closets

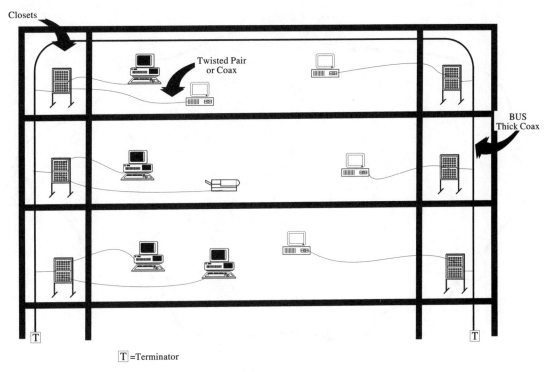

Twisted Pair
or Coax

BUS
Thick Coax

T

T

T =Terminator

Figure 1.10   The bus used as a hybrid with star configurations.

et (a physical star). This has some ease of installation and financial gains, but places all access in single closets, which could create single points of failure.

Token passing ring

In a token passing ring the topology is a ring. In this network the cable is laid out in a series of connections connected as point-to-point circuits, from device to device in a closed loop. Each connection uses a repeater to receive information on the inbound link, and repeats the signal onto an outbound link. As a more deterministic network, access control is handled via a token (permission slip) to allow transmission. The original ring topology was set up as described above and is shown in Fig. 1.11. This shows a wire-in, wire-out arrangement into the NIC card that provided the repeating function. As the network evolved problems began to arise that changed the way rings are installed.

In Fig. 1.11 a new station (Z) was required between devices C and D. In order to install station Z, the connections between C and D were detached. This brought the network down since messages could not be passed along the cable. This downtime can be potentially significant in a dynamic network environment where constant additions or moves are required. Obviously, something had to change. To overcome this problem of downtime in this envi-

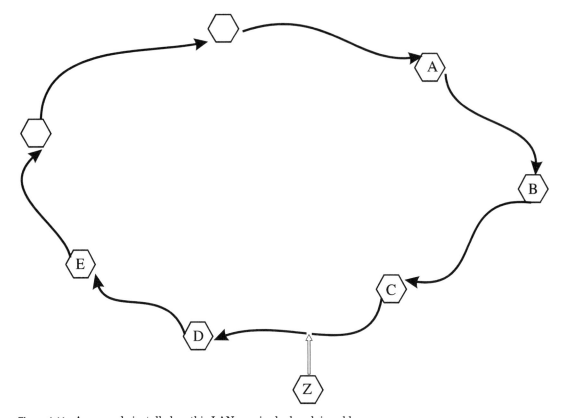

Figure 1.11   A new node installed on this LAN required a break in cable.

ronment, a wiring closet approach was introduced. In Fig. 1.12 the stations are wired back to a closet, where a MAU is added.

The MAU provides move, add, and change capabilities without the downtime mentioned above. If a device is detached from the network, a shunt bar (relay) drops into the loop. This allows the network to continue running, bypassing the downed station. However, the use of a MAU at a wiring closet introduces a single point of failure into the network. Note also that with the MAU a physical star is the topology being used; however, the LAN operates as a logical ring.

Token passing bus

The token passing bus is a hybrid of a bus topology and a token passing contention control method. Normally this network is found in a manufacturing environment, as defined by the manufacturing assisted protocol (MAP), the technical office protocol (TOP), or Datapoint Corporation's Arcnet.

A variation of the token passing bus is that it operates on a broadband cable system, where the others mentioned above were primarily on a base-

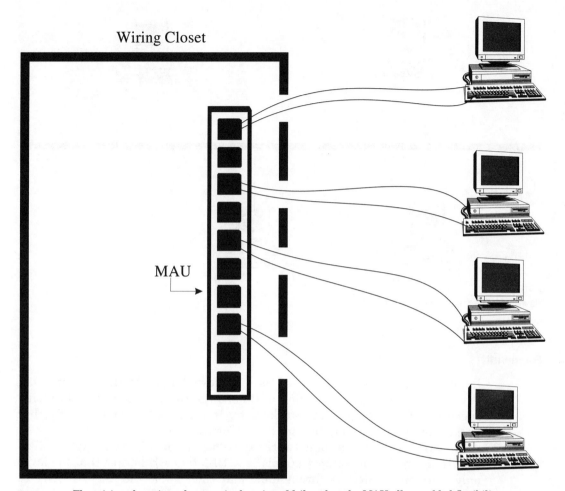

Figure 1.12   The wiring closet introduces a single point of failure but the MAU allows added flexibility.

band technology.* Figure 1.13 is a graphic of the token passing bus topology. In this scenario a bus is used, but control on the network is passed via a token (permission slip) in a logical sequence such as a ring. Since a single-node failure could cause the network to cease operation, a primary and secondary token passing arrangement is used. This limits some of the risks, but not all. The network shown in Fig. 1.13 passes logical control in a ring format. Device A passes control to device C; C passes its token to D; D passes to B, etc. In the event a node fails the secondary device will take control, bypassing the failed node.

---

*The discussion on baseband versus broadband is contained in the next section, entitled Bandwidth.

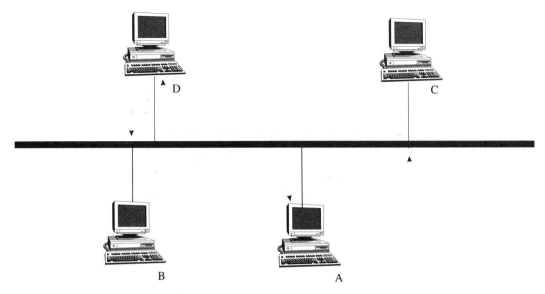

Figure 1.13 The token passing bus.

Bandwidth

Bandwidth on a LAN is the speed used to move data along the medium from device to device. One could think of this as the throughput or the operating speed of all the components working together. It is the capacity of either packets or bits per second transmitted along the cable.

In many cases we hear of networks that "bog down" when only 40% utilization is reached. An example of this is a 10-Mbits/s Ethernet that achieves only a 4-Mbits/s effective throughput. Why then would the LAN capacity be classified so high if only 40 percent can be achieved? The reasons depend on the number of devices, the size and types of information, the access methods to the cable, and the effective throughput.

For example, many LANs are built to support word processing information consisting of print files, words, and possibly a few other applications. Printers for example work at 9.6-kbits/s output/input with a high end at 19.2 kbits/s. Therefore to use the printer effectively a higher speed card must be added. This card will have its own memory buffers and processor. Each of the topologies or types discussed above is built around a set of standards. The IEEE 802 series describes the topology, bandwidth, access method, and recommended throughput. See Table 1.2 for a summary of these standards.

Although different in the topology, Table 1.2 represents standard bandwidths that are evolving for the types of LANs commercially available.

Baseband and broadband differences deal with the way the medium is used. To clarify this, an overview of these two techniques is in order.

TABLE 1.2   Summary of LAN Standards

Topology	Band	Bandwidth, Mbits/s	IEEE standard	Access convention
Bus	Baseband	2 or 10	802.3	CSMA/CD
Ring	Baseband	4 or 16	802.5	Token ring
Bus–tree	Broadband	2, 20, or 100	802.4	Token bus
Bus–ring	Fiber–baseband	10, 16, or 100	802.6	FDDI*
Ring (dual)	Fiber	50, 100, or 150	802.6	MAN†

*FDDI denotes fiber-distributed data interface.
†MAN denotes metropolitan area network.

Baseband

Baseband is the direct input of an electrical signal onto a cable, in its digital form. No modulation technique is used to place data onto the cable. Only one device at a time is allowed to use the cable, which is a time-division multiplexing technique. A media access control function is required to ensure that data is not corrupted due to multiple devices attempting to transmit simultaneously. In baseband systems, the signal is propagated along the cable until it is recognized by the sender as being delivered. Figure 1.14 is a representation of how a baseband system works. The typical networks running in baseband mode are the bus and ring. Once delivery of information from a device has been verified, the device relinquishes control for the next user.

Broadband

Broadband, typically used with coaxial cable, has multiple channels allocated for simultaneous use. It operates in frequency-division multiplexing form, and a separate channel is used for each frequency. In a coaxial world (community antenna television, or CATV) the data is modulated onto a radio frequency spectrum. This implies an analog form of transmission, although the input can be either analog or digital. Multiple devices can be transmitting on the

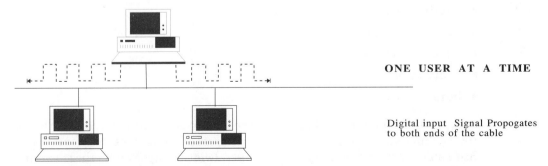

ONE USER AT A TIME

Digital input Signal Propogates
to both ends of the cable

Figure 1.14 Baseband allows one user at a time on the cable.

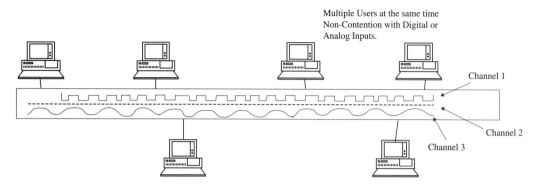

Figure 1.15 A broadband cable allows multiple simultaneous channel on a single cable.

cable simultaneously, on different frequencies. See Fig. 1.15 where multiple channels are being used on a single cable simultaneously. Usually each channel has a frequency band of 6 MHz, which would allow a user to run multiple 4- or 10-Mbits/s LANs on a cable. Most users do not use this technique; however, newer capabilities exist with fiber optics that use much higher bandwidth but typically use a baseband input. LANs that use a 100-Mbits/s capacity such as fiber distributed data interface (FDDI) use baseband input into a convertor that changes electrical signals into light (photonic) sources.

Switched multimegabit data services (SMDS) will eventually carry voice, data, video, and LAN traffic. In the future the services offered will allow much higher transmission than today. It will be important to keep everything in perspective. The greater the bandwidth, the greater the data rate. The higher the data rate the more loss in the event of a failure. This leads to a more pressing need to protect the information.

Media Used for LANs

LANs run on various media types; these include the following:

- Coaxial (baseband or broadband) cable
- Twisted pair cable
- Fiber optic cable
- Radio waves
- Infrared waves

Depending on the type of LAN and the bandwidth required, the selection of the medium varies. Figure 1.16 is a representation of the cable types used. See Table 1.2 for a summary of typical bandwidth supported by these various options.

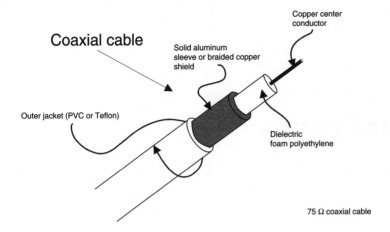

Coaxial cable

Copper center conductor

Solid aluminum sleeve or braided copper shield

Outer jacket (PVC or Teflon)

Dielectric foam polyethylene

75 Ω coaxial cable

Fiber optic cable

Strengthening member

Silicone protective coating

Fiber core

Outer jacket

Glass cladding 126 μm

Buffer jacket

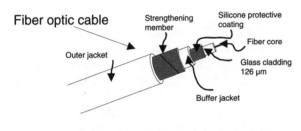

Single mode fiber cable

Buffer jacket

Fiber core

Glass cladding

Light source path

Cladding

Core

Outer jacket (PVC or Teflon)

Multimode fiber cable

Buffer jacket

Fiber core

Glass cladding

Light source path

Core

Cladding

Outer jacket (PVC or Teflon)

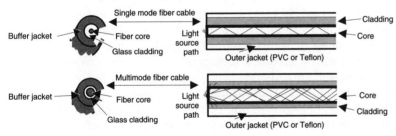

Twisted pair cable

Pure copper

Figure 1.16 Cable types include twisted pair, coaxial, and fiber.

Chapter

2

The Planning Process

Planning for Disaster Recovery

A startling statistic regarding the use of LANs has recently been published in various trade magazines, and studies from various research houses have validated the results. What these organizations have concluded on an industry-wide basis is that the average LAN fails approximately 30 times per year, or $2\frac{1}{2}$ times per month.

Regardless of the applications running on the LAN, this statistic indicates that network administrators must be prepared to recover the LAN this often! When first installed most LANs were designed around office automation services. These included applications for the delivery of word processing to the desktop, database access, graphics or desktop publishing, and electronic mail. Additional applications included facsimile, communications via modem pools, shared printers, and spreadsheets.

However, as the LAN movement caught on, users began to use the workstation as a means to bring more processing power to the desktop. Mission critical applications are moving from the host system to the LAN (*mission critical* applications are those that support the business function, and loss thereof will result in monetary loss or information loss). This means that host access via 327X (protocol or controller) through a 37XX front-end processor, for payroll and accounting functions, were delivered to the departmental level, leaving the mainframe as a large application server. Although this may appear plausible, it points to the need for protecting those PC applications that are more than ordinary PC applications, since host applications are now involved.

An example of applications that were accessed from a LAN terminal are shown in Fig. 2.1. In this scenario the host system (mainframe) becomes a large server on the LAN. A user on the LAN connects to the server and runs the application. This could be a very useful service to the LAN population. However, the LAN is somewhat fragile; this could cause major disruptions to accessing the host if the LAN fails. A user who makes a mistake and reboots their device (workstation) could cause the LAN to seize, thereby denying

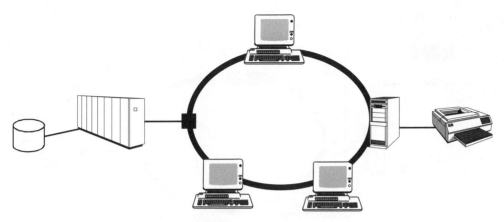

Figure 2.1   The host acting as a server on a LAN.

access to others using the host from the LAN. Since the data processing function may be in different hands than the LAN administration, several finger-pointing sessions may occur while blame is being assigned. The host as a node on the LAN can therefore be at greater risk of access failure based on what occurs on the external devices attached to the node. A virtual telecommunication access method (VTAM) application could continually fail if the LAN disrupts its operation. Furthermore, the loss of devices attached to a VTAM application could cause runaway processes. Although the SNA environment is isolated from the LAN itself, the loss of a device running the application could cause problems. [Systems Network Architecture (SNA) is an IBM mainframe plan. Using a synchronous data link control (SDLC), IBM introduces the 3270 protocol and 327X devices (i.e., terminals, controllers, etc.).] This is not a critical problem, but leaves the door open for the LAN to disrupt multiple users on the host, the controller, or front-end processor attached to the LAN. Again the concern here is the impact on host applications when the LAN fails, especially since other direct-connect terminals to the controller will also likely be affected.

In Fig. 2.2 a similar scenario is shown but a front-end processor (FEP) (37XX) is used as the node on the network. This isn't much different in terms of exposure. The FEP is designed to isolate the host from failures on the LAN. The network control program (NCP) residing in the FEP would likely fail if the LAN fails, causing the NCP to restart. Although isolation for the host exists, other devices, terminals, and printers will be disrupted. This involves all of the devices attached to the host through the front end, leaving the host functioning but the rest of the network down. What would be the gain of isolation if everything else still fails?

In a third possible scenario the FEP is still attached to the LAN, but a 3174 cluster controller is added as shown in Fig. 2.3. The 3174 controller is a node on this token ring network. With a token interface card (TIC), logical sessions are established through the controller for the PC's running 317X

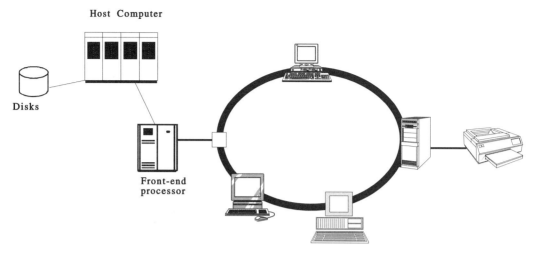

Figure 2.2 The front-end processor attached to the LAN provides isolation to the host.

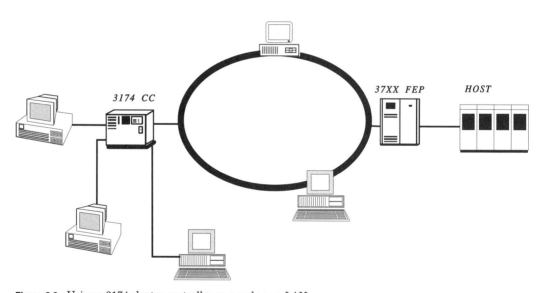

Figure 2.3 Using a 3174 cluster controller as a node on a LAN.

emulation. If the LAN fails, the sessions on the 3174 controller will be lost and the FEP will drop all sessions. Although the host is still running, all connections are lost. Again, the problem is the same; we expect the host to be constantly available, but the LAN is more prone to failure. A single point of failure causes major disruptions on the mission critical applications whether on the host or ported down to a server on the LAN.

The fourth example depicts the use of a 3274 cluster controller attached to the LAN. As shown in Fig. 2.4 the 3274 controller with a TIC is merely a

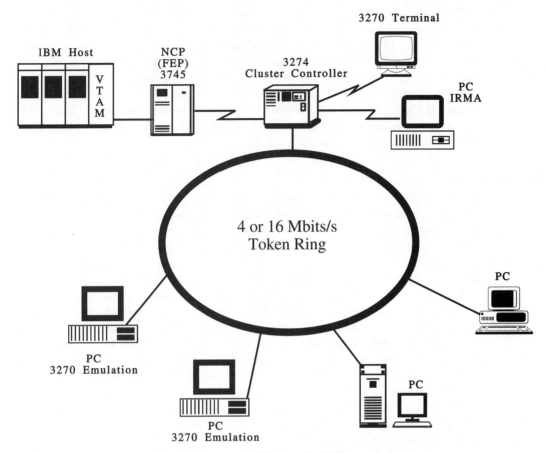

Figure 2.4   The cluster controller allows access to SNA from the LAN.

gateway function to the SNA environment for users on the LAN. Once the LAN fails the 3274 controller goes down, impacting 3278/3279 terminals or PCs with 3270 emulation (an IRMA card). [IRMA is a 3270 emulator card that makes a PC look like a 327X device (3278/3279 terminal). It is a specific vendor product from DCA.] All these devices may be affected even though they are isolated from the LAN on the controller.

In each of the above cases, companies have migrated to LANs as either the primary application processors or as an access method to a host. Even though these examples seem to be extreme, the reality of these situations is true. More users are demanding host access through their LANs. They have found that even though local processing is important, they still need to run the applications running on the host systems in the data processing department. This exposes the once secure host world to greater risk because of the unplanned and uncontrolled external connections from the unprotected and unreliable LAN. Because of this movement, the conflict between data processing and LAN administrators has dramatically increased.

Thus, the more critical applications began to work their way out of the controlled environment. They landed in the users' hands, which put the onus to protect access to the data on the department manager or a network administrator. Unfortunately these new responsibilities are being given to operational people who have not been trained in the fine art of protecting the systems or the files. When a network fails, it is too late to wonder how to recover. Actions toward the preparation for problems must begin before they occur. In reality, planning for disaster recovery should begin when the LAN is being planned or designed. However, this rarely happens.

The need to plan

As the scenarios highlighted above all come into play within an organization, the need to plan for the inevitable disaster becomes more crucial. Since the LAN may be under the control of non–data processing personnel, the long-term impact to the organization may not be as evident. If the LAN fails, just what should be done first to recover it? Many LAN administrators have spent hours just trying to get a handle on what is actually going on after a failure. For the most part, the administration of the LAN is an on-the-job training situation. The proper tools and methodologies to conduct appropriate troubleshooting may not be available. If the LAN supplier has not shown the need for these tools, then they probably will not be on hand. All of the other issues mentioned in Chap. 1 may also be occurring (i.e., the lack of controls, the attitude toward data processing staff). Consequently, the need to start the planning process may not be as evident as it should be.

Initially, a LAN administrator should take a very close look at the topology and the pieces installed on the LAN. A "new vision" approach should be used, as it is all too easy to look past potential risks, since the administrator is also the architect of the LAN. Normally the end user will ask tough questions of the MIS staff, for example:

- How is data protected?
- What procedures would be used to obtain recovery from a failed component?
- How can data be retrieved from archival systems?

Yet when building a LAN for themselves, these questions are often unasked and therefore unanswered. The supplier of the network should provide the answers.

Figure 2.5 is a summary of critical issues that should be addressed. More specific questions can be added to this list, based on the individual organization's situation.

What if the LAN fails?

The list of questions contained in Fig. 2.5, although not all inclusive, should portray a feeling for the degree of complexity that may have to be dealt with. If the LAN failed *right now*, who would know what to do first? Is there only one

Questions to Ask the Vendor or Yourself

1. If the LAN fails, where will the support come from to get it back up?
 	
   a. Do they have the proper personnel?
   b. Do they know what has to be done?
   c. How often do they get the chance to recover a network?

2. What business operations will be impaired as a result of failures?
 	
   a. Are critical operations run on the LAN?
   b. Do other alternatives exist for critical users?
   c. How would you migrate these applications onto another server, workstation, or LAN?
   d. Do these resources exist?

3. What would the organization lose if access to the LAN is denied?
 	
   a. Would this expose the organization to ridicule in the industry?
   b. Would the loss of the LAN jeopardize any security issues?
   c. Would production stop, rendering several or hundreds of employees unproductive?
   d. Is management support at risk if the LAN fails?

4. What losses could the organization sustain, if the LAN fails?
 	
   a. Financially, will the organization's position be unnecessarily burdened?
   b. Could the company's competitive position be impacted?
   c. Are there legal risks associated with downtime?
   d. What is the maximum loss the organization could sustain?
   e. Can this loss be minimized through other techniques?

5. Who knows how the network operates?
 	
   a. Do others know how the LAN is laid out and how to recover it?
   b. If you are not available, who will do the work?
   c. Who creates the backups? How often are they done? Where are they kept?
   d. Do any others know how to use a backup to recover information?

6. If the LAN fails, which applications should you recover first?
 	
   a. Who sets the priorities?
   b. How do you maintain these priorities?
   c. Do users know this process, so they will be patient? Or will chaos break out with every user demanding service immediately?

7. If you lost the use of your building, what would you do?
 	
   a. Can you get space easily?
   b. Can you get properly configured equipment quickly?
   c. Are the backup files and/or tapes available off-site?
   d. Who is going to do all the work?
   e. What will all this cost?

8. Do you have a budget for recovery?
 	
   a. Who controls this budget?
   b. How much is allocated?
   c. Will the budget cover any scenario?
   d. Does management support this budget area?

Figure 2.5 Critical questions that should be answered when constructing a LAN.

person in the organization who knows what to do? Unfortunately, this is the case in many companies. A single person has all of the information stored away.

Some organizations have decided that only one person is required to support and recover a LAN. Although this may be fine for day-to-day events, it is unfair to expect only one person to be totally responsible. The company cannot expect the individual to be available seven days a week, 24 hours a day. This is an easy path to disaster if this is the company's position.

Thus, when looking at the environment a critical look must be given to all the pieces on the LAN including the human resources side of the equation. The intent here is not to create conflict with budgetary constraints or hiring restrictions. The opportunity exists to recruit from within the company on a part-time basis. Departments whose operations will be impacted negatively will likely offer some assistance in assuring a quick and orderly recovery. If a department manager has 26 users on a LAN that fails, the entire staff becomes nonproductive. By having a designated individual trained and ready to help in the event of failure, the network could be recovered more quickly. Thus, the 26 people become productive again, which benefits the department, as well as the whole organization. So use this concept wisely.

If any doubts exist that the need for a recovery plan is obvious, think again. To better understand this, we could define disaster recovery for LANs as follows:

> *Disaster recovery:* The organization's ability to get back into business quickly after an event that disrupts the flow of information. This is done through a set of preplanned, coordinated, and totally familiar procedures with an established set of priorities.

Stated a bit simpler, disaster recovery for LANs is an orchestrated set of policies and procedures to get the employees back to work quickly. This is no different than the data processing disaster recovery planning process.

The goals of any disaster recovery planning include:

1. Minimize the disruption to the bare minimum.
2. Effect the return to normalcy as quickly as possible.
3. Maintain control over a series of events where chaos will reign.
4. Prioritize the efforts of people, the sequence of events, and the strategies used during the recovery and restoration phases.
5. Ensure the security of information and equipment so that other consequential disasters will not occur.
6. Gain and maintain management support and confidence.
7. Prevent other consequential damages as a result of this disaster.

Each of these goals is attainable, but if done after the fact could take significantly longer. By preparing for the inevitable failures and disasters in advance, LAN managers earn the respect and confidence of many, including management. Additionally, by having a prepared set of procedures, the chances of omitting any procedures are lessened dramatically. People under duress have a propensity to shoot from the hip, making poor decisions. Others who have no plan freeze, making no decisions and taking no action. In either case the result can cause more damage to the organization than any other action.

Planning strategies

What is the best way of introducing this planning concept to management? How might the process be expedited to gain support, funding, and cooperation at all levels? There is no one simple answer. Every organization is unique. Various approaches are available. Depending on the organization these can either be effective or they could backfire. The options are as follows.

1. *"Chicken Little" strategy:* Walk around constantly preaching gloom and doom, trying to get people to wake up. However, this could *also* create the "cry wolf" syndrome, where management ignores future cries and pleas for help. As the administrator or manager of the LAN, creating a negative image could result in doubts about your ability to handle the network.

2. *Ostrich management:* By hiding behind other activities much like the ostrich buries its head in the sand, you can pretend that nothing can go wrong. Of course, then, when things do go wrong, play the Chicken Little or the finger-pointing routine (listed below). Surprisingly, this approach is one that promulgates throughout entire organizations. It also is the one most widely used.

3. *Finger-pointing strategy:* Why be concerned when others can be used to recognize a problem? When something goes wrong, blame them. It is their fault for not taking the proper precautions to protect and recover their network and data. This technique may work once, but it is not sound strategy for any business. Many organizations rely on this technique up and down the echelons of management.

4. *"Superhuman" strategy:* Any LAN administrator knows that superhuman efforts are needed to recover a network after a disaster. Biding time until a failure or disaster occurs, then coming to the rescue and saving the organization may earn extra respect and confidence. However, the superhuman effort may go unnoticed or cause other concerns with management. "Why wasn't someone else already trained to help?" is not an unrealistic or uncommon question. What if the LAN manager gets into an accident on the way to work? Who covers the LAN then? Where does this leave the organization's ability to get back into business quickly?

5. *"Close to the vest" strategy:* Do not tell anybody anything! Keep all information hidden or memorized in your head. Never write anything down. This will not guarantee a long-term career, however. Many managers or administrators have good intentions. They recognize the need for complete documentation but never get the time to complete it. Consequently they must always be available to recover this network.

6. *Business team player strategy:* Start recruiting assistance from concerned users and managers alike. Build a team of knowledgeable people who can spring into action. Sell the concept to upper management; do not expect them to ask for the protection. Realize that things *"can and will go wrong"*

and prepare for it. Be business oriented, not technologically oriented in assessing risk and planning the strategies. This approach will go a long way in earning management's respect and support. Users will also feel more comfortable since they are active participants in the process.

Decide which strategy works best in the organization or develop a hybrid of two or more into a single plan. Feel free to use these as you will.

Management Commitment

Obviously the cost of preparing a recovery plan with all of the associated training, documentation, hardware and software backups, and the time to produce the plan itself can be significant. The cost depends on the size, the degree of LAN activity, and the level of management's commitment.

What management will see is a large "sink" against either capital or operating funds to protect the very device that is supposed to reap huge savings. A throwaway mentality from vendors, users, and management can get in the way: "If a device fails, throw it away and get a new one." But what of the data and application side of that device? Can the company afford not to have a plan to recover the critical information necessary to run the business? Many organizations feel that they have been sold a bill of goods. LANs were supposed to be the inexpensive solution to runaway data processing costs. Now protecting the LAN requires additional personnel and equipment. This could leave management with a soured attitude towards the value of the LAN.

Some organizations have stated that they will not spend any more to protect the LAN than the cost to replace the hardware. This is a mistake because the hardware is the least expensive part of the LAN. Productivity and critical information far exceed the cost of the LAN hardware. If this approach goes unchecked, then true disasters are lurking in the environment. Be careful not to fall into this trap; it could be devastating to the company.

Prepare to meet management head on. Compare the cost of doing nothing versus the cost of full-fledged prevention and recovery planning. Ask members of management what the cost would be if the LAN fails for two days, one week, etc. Do they know? Does anyone know?

Justifying the plan to management

Obviously when approaching management with this issue, expect some resistance. After all, they have been told that the LAN will save them money and that the use of a LAN would offset the exorbitant costs of the mainframe environment.

As novices to the protection and recovery planning process, LAN administrators bought only the components necessary to equip the departmental user with single systems. In many cases, to keep the prices down, servers are used to suit dual purposes (workstation and server combined). Now they will attempt to sell management on the need for additional hardware and software to provide the minimal amount of protection. They can expect senior management to be less than enthusiastic with this new requirement to fund a recovery and protection process. The name of the game here is money! How much protection must be provided, and at what price?

Figure 2.6 provides a list of possible areas to be considered when justifying disaster recovery planning to management. These issues will address the true need for the plan and the preventative measures on the LAN.

First and foremost, the hardware issues have to be cast aside. Instead, prepare to meet management on a business basis. Consider the cost to the organization if the data is lost.

As a result, the need to apprise management of both the risks to the LAN as well as to the business impact due to losses should be stressed. To do this a business impact analysis would assist management in evaluating the funding and resource commitments necessary. This business impact analysis would include the following:

1. The list of outage events

2. The probability of the event occurring

3. The length of outage (average)

4. The cost of the outage (time, money, other)

5. The prevention steps that could be taken

6. Resultant length of outages with procedures

7. The cost of prevention and/or recovery

8. The differences

Figure 2.7 is a form that can be used to represent these issues. By comparing the two options, with or without an action plan, management can select the best mix of strategies available. The form can be filled out line by line with different risks and probabilities that the risks will occur. Using this risk assessment and putting tools in place to prevent the risk from occurring will yield differences in costs.

An example of this might be as follows:

> An event such as a major disk crash on a file server could occur. If this happens, access to all files will be denied for a period of 1 to 2 days to 26 users associated with the LAN. The average hourly rate for the users is $20.00 and the probabili-

---

Areas to Consider in Justifying LAN Disaster Recovery Planning

❑ Cost of lost revenue

❑ Cost of lost productivity

❑ New hardware and/or software

❑ Customer and employee confidence

❑ Marketing efforts to overcome an image problem

❑ Contractual obligations not met may yield consequential losses

❑ Legally the company may be bound or required to have a plan in place

❑ Other reasons that are particular to the organization

Figure 2.6 Justifying disaster recovery planning; consideration points.

JUSTIFYING DISASTER RECOVERY PLANNING BUSINESS IMPACT ANALYSIS:									
NO ACTION PLAN				WITH ACTION PLAN					
Event	Probability	Length of outage	Cost of loss $	Probability	Length of outage	Cost to recover $	One time cost $	Priority High/Medium/Low (H/M/L)	Time to implement

Figure 2.7   The business impact analysis worksheet compares the risk with and without a plan in place.

ty of the event (historically) is 0.7, with 10 occurrences on average per year. Expanding this we can say that the difference will be between $25–55K by making a simple investment of $2,500. This is shown in Table 2.1.

These numbers are variable, since the 0.7 probability would be 1.0 if the event occurs.

Expanding this further, the other side of the equation is what effort and costs would apply in the preventative mode. Using a duplexed or mirrored

TABLE 2.1   Cost Comparisons of a Backup Disk Protecting a Hard Disk*

	With backup	Without backup
Initial cost	$2,500	...
Lost time	$1,040	$29,120–58,240
Total cost	$3,540	$29,120–58,240
	Savings $25,580–$54,700	

*The numbers shown are strictly estimates.

disk approach the average downtime may be 0 to 1 hour. *Assuming* a one-time cost of $2,500 for this backup system, the results would be as follows:

Downtime = 1 hour

Average hourly rate per user = $20/hour

Number of users = 26

Probability of event = 0.2 (dropped since backup system is in place)

Number of events per year = 10

[(1 hour) × ($20/hour) × 26] × 0.2 × 10 = $520 × 0.2 × 10 = $1,040

It would take only one event to pay for the shadow disk. The savings estimate is given in Table 2.1. Remember this cost is *only* for lost productivity! The other costs such as lost revenue and lost confidence, and consequential damages, would have to be calculated and inserted into the total equation.

The Preliminary Plan

In order to get support and financial backing from management, it is obvious that management will have to be convinced that there is a need for the planning process. Managers will also have to be convinced that there is a valid business need for the plan. This will take on many forms, not the least of which is a method of showing management what the impacts will be without the plan. So for now, consider the consequences. Many of today's managers are acutely aware of the situations that have transpired in industry (and no industry is immune) where the business entity (company, firm, supplier, or carrier) is in jeopardy of sustaining operations when disasters strike.

However, business managers have a tendency to describe these needs in nonbusiness terms (see Fig. 2.8 presenting the preliminary plan). Data processing or LAN terms are used extensively, which are comfortable in everyday discussions with vendors, yet to approach management using these terms is a problem. Managers will be somewhat unenthusiastic with technical requests. The primary reason is obvious to the layperson: Management neither understands what is being said, nor does it care about the technical terms being used. Management is far more comfortable with a business proposition. When using management's business and financial terms, there is a reckoning of the severity of the situation.

For example, tell management that without a plan in place, the company would risk a severe decrease in market share in the event of a disaster, or that the organization will only survive as an entity if a plan is in place to reestablish connectivity capabilities within 24 to 72 hours.

Management of any organization deals with business issues. The above examples will get management's attention. However, be prepared to answer

Figure 2.8   Presenting the preliminary plan.

any questions. If challenged, win management's support for the survivability issues. Here is where the networking portion of the planning process comes into play. Before you ever get an audience with management, work with key departments in the organization. Recruit their assistance in pulling information together for the presentation to management.

When presenting financial costs to management, more credibility is associated with numbers (dollar amounts) prepared by a finance than by a LAN manager. The LAN manager is perceived as the technical person, whereas the controller is perceived as the numbers person.

Recruit whatever resources are necessary to win management commitment and long-term support. Network inside the organization to get support for a disaster recovery plan. Recruit from finance, facilities, security, audit, human resources, etc., departments. Each of these is likely to be on the LAN and have a vested interest in providing protection.

The Presentation

Be upbeat, be positive, be confident and accept management's questions or interruptions as constructive. *Do* use these questions as opportunities to reinforce strong points, rather than letting them frustrate the effort. Many highly qualified, intelligent people fail when trying to deliver a program presentation to management. They suffer from fear of public speaking. Although they have put together an outstanding plan, they fall flat on their faces when they try to present it. As a result they are sent away with no decision, or a series of questions or added items to address. In fact they had the answers with them. This is unfortunate because they lose too much time and effort when this occurs. But, how do they get around this problem? If public speaking is uncomfortable, yet the plan must be presented to senior management, what can be done? Try to develop the necessary skills before going in front of the board or recruit someone else to do it (see Figs. 2.8 and 2.9). Here is where the assets such as the key people discussed above (i.e., public relations or sales) can present either all or part of the plan. They are very qualified to do it. However, a word of caution is in order: make sure to retain ownership of the program. Often times when someone else does the presentation, he or she will start to feel a stronger commitment and begin to take ownership for it. Use the resources to your advantage.

So what exactly will be presented to management? Some ideas to consider are listed in Fig. 2.10.

Remember:

1. Keep it simple and short (the KISS method).
2. Talk to board members in *their* terms, not in LAN terms.
3. Use business terms that they are familiar with.
4. Do *not* use technical terms.

Figure 2.9   Remember to use these points while delivering the presentation.

1. Introduction: just why management is here.
2. Historical facts: who has been affected by LAN disasters; how have they survived? Use examples from the same industry, where possible.
3. Risks and exposures: be frank and explain specific company risks as opposed to industry risks.
4. Impact: what happens to the business if a disaster strikes?
5. Prevention of major outages: a preferred method is to prevent disasters rather than have to recover from them.
6. Steps to be taken to develop a plan: this will include any options available.
7. Resources and staffing needed: this includes both the development of the plan and the ongoing maintenance and testing.
8. Estimated cost to produce the plan: the cost has to be all inclusive, for human resources; options (consultants, programs, awareness), etc.
9. Length of time necessary to put the plan together: this has to be realistic in terms of putting the plan together; make sure to stay on schedule wherever possible.
10. Possible legal liabilities that will exist if a plan is not in place: this covers several legal aspects, such as contractual, industry requirements, shareholders, etc.

Figure 2.10 Organizing delivery to management.

Presenting the plan to management

1. Introduction. Here is an opportunity to address the key concept of disaster recovery and restoration for local area networks. Have an estimate of just what you are spending annually for LAN connectivity services. Try to put together an estimate of revenues derived from this investment. If possible show a direct relationship of LAN dollars spent to revenues or profits derived; this is a good basis to prove to show management just what it is that they are trying to protect.

2. *Historical facts.* How many companies have suffered from disasters? There are a few listed in this book, but there are hundreds that have never been heard of. Do some research in your industry, geographic area, etc. Next, try to put some dollar value to other companies' losses, from a historical perspective. If any companies have folded as a direct result of a disaster and could not sustain the losses, use that information in your presentation.

There were over 40 major natural disasters last year alone, from the earthquakes in Southern California, floods in the midwest and southwest, hurricanes in the southeast, severe cold (yes, this can be classified as a disaster), tornados, lightning strikes, and others. Added to these natural causes are the human errors that caused various types of disasters: fire, floods from broken pipes, power outages, viruses, hackers, bomb threats, looting, rioting, and terrorism, cut cables. This historical background can have a very strong impact.

3. *Exposures or risks.* This is a more specific topic: What areas are exposed in the current environment? Try to be realistic; *do not use scare tactics* here. Present information using a matter-of-fact approach. Identify all of the weak points. In many cases this subject will become far more detailed when going

beyond the initial presentation and beginning a complete inventory. If possible, provide a global overview to the managers, they will have an appreciation of their needs and be more receptive to funding the planning process.

There are some additional items to consider when addressing management, such as the following.

- *Contracts* that can be set up to help in the prevention mode with the various suppliers and vendors that provide services and equipment. Many of these suppliers will contractually agree to meet an organization's recovery needs within a prescribed time (48 to 72 hours).

- *Policy and procedures* changes necessary to implement the plan. These may affect day-to-day operations, personnel policies, and the like.

- *Legal liabilities* from both the organizational level and from a contractual level. With the "just-in-time manufacturing" concept, the liability that passes on to the incidental suppliers of a product or service can be significant. Does the company have any contracts to deliver products and services to another organization that could jeopardize either organization's position of a financial or competitive nature?

- *Insurance* requirements and costs that could be reduced if a plan was in place. Insurers charge premiums based on the risk of a payout and the dollar value for these payouts. If a LAN position is improved due to the presence of a recovery plan, can the premiums be reduced? If the premiums are reduced, will the money saved on insurance be moved into the disaster recovery budget?

- *Regulations* that may be in existence requiring a plan (based on the industry) or expose the organization to severe penalties or risks of shutdown if a plan is not in place.

Although these topics may seem beyond the scope of the LAN planning process, surprises (unpleasant) may occur if they are not considered in advance. All too often, LAN managers forget the basic rule of remembering what their business is and what their corporate charter is. The mission statement for the organization should be in support of the total organization and not limited to the computer and LAN functions. The company will be better served when this subtle change in thinking is remembered and internalized.

Do not forget to discuss the planned methodology to develop the plan. Management needs to understand that there will be some associated costs with putting the plan together. Options exist, of course. The differences among options are varied, and the related costs for each will be significant. Most companies look at developing a plan as a result because of management's inquiry about what is being done in this area. Assume that there is some awareness of the need for disaster recovery planning. It may have come from an audit report, a newspaper article, a trade magazine report, or just a discussion between executives. At any rate, decide to use the best proactive approach, depending on the time, resources, and money needed.

Options in Developing the Plan

1. You can attempt to develop the plan in-house with existing people and equipment. There is a cost associated with this: time and equipment costs. There is also a risk that the time to develop the plan just will not exist, because of daily responsibilities and dealing with crises that we all go through. Thus, this project starts to get pushed into the background of our minds. This can be flirting with danger, since you have stated that you will get the plan developed and with in-house staff. The cost to do this can range from approximately $25 to over $75 thousand depending on the resources and equipment you commit.

2. Since auditors may have brought the need for a plan to management's attention, you could suggest that the company use them to help develop the plan. The Big Eight accounting firms (or five if that is the number these days) are all starting to become very active in assisting companies develop these plans. Their fees are not inexpensive, but they do have management's ear and respect in most cases.

3. There are hundreds of independent consultants in the industry today (in excess of 600 firms, and growing steadily) who can be hired to work with you on the plan. Their expertise and costs are subject to individual review. However, be advised: do not let any consultant or other entity write your plan for you. You have to be an active participant in the research and development of the plan. This plan will become a "living and breathing document" that must be constantly tested and updated. It will not serve you or your organization to have some outsider write it for you. The costs for consultants can vary, but you can expect to spend from $1,000 to $1,500 per day, plus expenses for a good consultant. You have to decide how to best use your company's money.

4. There are recovery service vendors who can offer consulting, assistance in plan development, and even a "hot," "warm," or "cold" site as part of the plan. A "hot site" is a prepared site, treated with environmental protection (uninterruptible power supply) systems, fire detection, air conditioning, etc.), and all of the PCs and printers necessary for a company to move in and get things running quickly. A "warm site" is similar to a hot site, but will not have the LAN equipment there. A customer would deliver equipment and set it up. This takes longer but is less expensive. A "cold site" is an empty building that must be prepared before any action can take place. Recovery takes longer. There are obvious biases built into these organizations' planning processes; they are in the business of selling an insurance policy in the form of a contingency site. Therefore, it is only fair that you should expect them to recommend their own services. Costs for this assistance would be tied into their services performed and contingency site services subscribed to.

Developing your plan

"Do it yourself"

Advantages

1. *Control:* The planning and execution of the project rests in your hands.

2. *Environment:* You know your operations and requirements the best.

Disadvantages

1. *Time constraints:* pulled from crisis to crisis.
2. *Cost constraints:* $25,000–over $75,000.
3. *Resource constraints:*
   a. Do you have the staff?
   b. Do they know what has to be done?
   c. Do you have the tools necessary?

Auditors (e.g., the Big Eight)

Advantages

1. *Resources:* They can use as much as they need.
2. *Respect:* Senior management will listen readily to what they say.
3. *Intuition:* They can sense what may unfold.
4. *Skills:* They are very capable of presentations to senior management.

Disadvantages

1. *Ownership:* If they do it for you, can you accept ownership?
2. *Complexity:* Auditors are known for producing complex volumes.
3. *Cost:* Very expensive; prepare to pay $1,500 to $2,000 a day.
4. *Knowledge:* Auditing firms may have to acquire more technical knowledge to help with the plan.
5. *Updates:* Who does the constant updates of the plan if they write it?

Consultants

Advantages

1. *Resources:* Consultants get as many as needed.
2. *Timing:* They can help you get the job done quickly.
3. *Credibility:* Consultants may already have the acceptance of senior management.

Disadvantages

1. *Ownership:* Who will put it together? If they write it, can you accept ownership?
2. *Cost:* Plan to spend from $1,000 to $1,500 a day.
3. *Experience:* Most consultants come from a mainframe data processing background and may not have LAN experience.
4. *Updates:* Who will do the updates?

Chapter

3

Physical Protection

When reviewing the LAN, addressing the multitude of areas involving physical protection is a must. For the purposes of reviewing this area, physical protection will address the following items, as shown in Fig. 3.1. These areas deserve special attention.

Some of the largest potential risks exist in the areas listed in Fig. 3.1. The two most common problems involving LAN disruptions are in the cable and power areas. A little protection can go a long way. Protecting these areas is not a very expensive proposition, but unfortunately they are the most commonly overlooked areas and the ones taken for granted most often.

Cable Systems

Depending on the type of media used (the cable types were outlined in Chap. 1), the cables are prone to some very serious disruptions. From the basic twisted pair system (both unshielded and shielded), risks exist that subject the LAN to cable cuts or other damage that could severely impair the ability to transport high-speed data on the cable. These wires look like normal telephone wires. If they run in normal riser shafts and conduits throughout the office area, they could be mistakenly pulled, tugged, or cut by anyone working in and around the office area. Maintenance personnel may cause disruptions to the LAN when working in plenum or conduit spaces. Many new runs in existing conduits could literally strip off all the insulation from an existing cable or multiple cables. This can cause devastating results on the LAN. Network

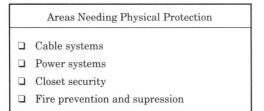

Figure 3.1 Major areas of physical protection.

cabling constitutes a significant amount of network downtime (greater than 50 percent). Improper cabling or poor installation make it extremely difficult to maintain any network environment. Take the time and effort to do it right the first time. This will prevent long-term nightmares in the future.

The best way to ensure high availability is to begin correctly. Planning is extremely important. Make sure that cables are properly designed and installed. The following guidelines will help in assessing the cable situation. From the narrative listed below, Fig. 3.2 is a summary or checklist of the areas covered.

1. Keep the proper distances from power sources such as motors, fluorescent lighting fixtures, or any other sources of interference. Minimum clearance should include the following as a rule:

   a. For twisted pair wires maintain at least 6 to 12 inches from high-voltage lines, transformers in electrical closets, etc.
   b. For coaxial cables, keep a minimum clearance of 4 to 6 inches from power lines and transformers.
   c. For fiber-optic lines, no clearance rules apply, since fiber is not prone to electromagnetic or radio-frequency interferences (EMI, RFI).

2. Use only high-quality, high-grade cables. If using shielded twisted pair (STP) cables make sure the proper grounding procedures are used. Depending on the speed of the LAN, buy the best cable for present and future needs. This will prevent future problems. The cabling systems specifications have been developed based on speed and distances. These standards are based on levels of throughput, as shown in Table 3.1.

   These standards are based on 24-gauge twisted pair wires and fit well within the throughput of the cable. Normal twists of the cable will run at about 20 per running foot of cable. Since the copper acts as an antenna, higher throughput on the cable could produce serious EMI inductance fields, thereby jeopardizing the integrity of the data. For high-speed data a twist rate of 40 twists per foot is recommended. This includes the jumper or patch cable in wiring closets, where typically flat wire is used.

   For cable used in an IBM wiring system, IBM developed types of cable to meet user needs. These specialized cables fit very specific functionalities and are shielded twisted pairs. These standards (from IBM) are given in Table 3.2. Because of the shielding, many LAN users have a propensity to keep away from this cable system. Shielding can cause far more problems than the benefits to be gained.

   The cost to increase the capacity (or thickness) of a wiring system is far less expensive when installing it the first time than having to go back and rewire later. Therefore it makes sense to use the best grade of cables from the beginning.
3. Terminate the wires in clearly marked patch panels, make sure the cable runs are labeled on both ends.
4. Document every run thoroughly and test it to ensure accuracy.

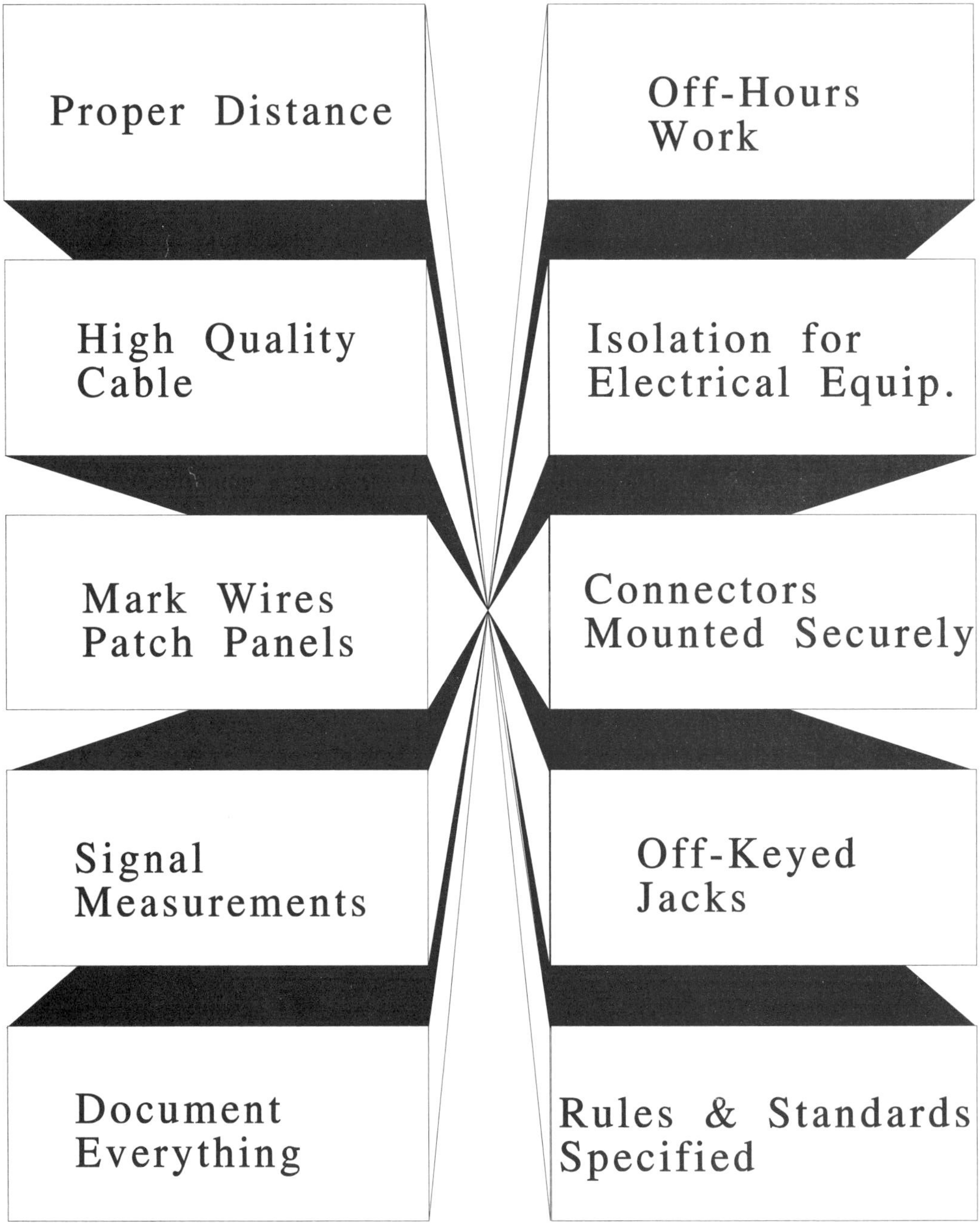

Proper Distance	Off-Hours Work
High Quality Cable	Isolation for Electrical Equip.
Mark Wires Patch Panels	Connectors Mounted Securely
Signal Measurements	Off-Keyed Jacks
Document Everything	Rules & Standards Specified

Figure 3.2 Checklist of wiring systems inspection.

TABLE 3.1   Standards Set to Ensure that Proper Levels of Cable Are Used Based on the LAN Speed

	Standards Based on Levels of Throughput (Shielded and Unshielded Twisted Pair Cables)
Level	Description
1	Telephone service, such as analog and low-speed digital
2	EIA-232 specifications at up to 400 Mbits/s
3	10 Mbits/s, <100 m
4	16 Mbits/s, ≈100 m
5	100 Mbits/s, ≈100 m

TABLE 3.2   The IBM Standards by Wiring Type

	Standards for Use in an IBM Wiring System
Type	Description
1	Two pair shielded 22-gauge wire (solid)
2	Two pair shielded 22 gauge plus four pair unshielded 22 gauge (solid copper)
3	Four pairs of unshielded 22 gauge (solid)
5	Fiber optic (multimode)
6	Two pair shielded 26 gauge (stranded)
8	Flat cable for under carpet installation
9	Two pair of shielded 26 gauge (solid)

5. Conduct signal measurements after any work is done on the cable system. Time domain reflectometer (TDR) measurements will show any open circuits, short circuits, or grounds on the line. A TDR can be a separate system rented for short periods of time to test the cable plant. Software systems now offer the capability to "look down" the cables for several factors that can adversely impact the operation of the LAN. Some of those software solutions could include the ability to isolate or locate the items listed in Fig. 3.3.

6. Only allow work during noncritical periods. Schedule work during off-

Problems on Cabling Systems
❏ Open circuits
❏ Short circuits
❏ Bad taps
❏ Splices
❏ Bad bending radius along the cable system
❏ Grounds

Figure 3.3   TDR or software solutions can locate these problems.

hours and have a technical person around to monitor the work being done. Make sure that a physical inspection is done to ensure that damage to cables has not occurred. This is particularly true when new cables are installed in conduits or plenums. The installers may well have damaged the wiring by stripping the insulation from the wire, causing open copper to be exposed. This will create an "antenna effect." Another area that will be at risk is at the jacks (modular connectors), where tugging and pulling on cables can cause the copper connection to pull away from the connector pins in the jack. This can lead to intermittent problems, which are harder to isolate and resolve.

7. Provide isolation for all repeaters and electronic hubs in the closets away from electric transformers, motors, etc. Use separate, solid earth grounds, keep minimal clearances, and where possible provide power isolation transformers.

8. Connectors at the workstation should be securely mounted on the wall or on furniture. Do not allow them to be loosely draped across floors or dangle under the furniture. The user will roll over these with a chair, crimping or stripping the insulation of the cable, causing more risk and damage.

9. An optional "off-keyed" jack or block will prevent users from plugging in other devices that do not belong in this particular jack. The major risk is a plug inserted into the jack that introduces high voltages, which could destroy the electronic equipment at the ends of the cable. Figure 3.4 is a sample of the off-keyed jack.

10. Specific rules or standards must be set, stating that no one can add a device to the network without being certified. This limits a good deal of the risk, if it can be enforced.

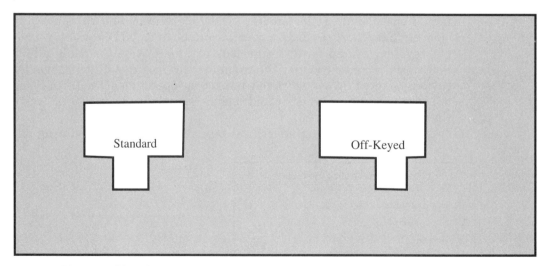

Figure 3.4 Using an off-keyed jack prevents other devices from being plugged into the LAN.

11. An alternative to item 10 is to allow only members from LAN management to unplug or move a device.

All work being done within an area that could subject the wiring to exposure or disruption should be coordinated with the LAN administrator.

Power Systems

One of the problems that will usually be overlooked in an office environment is the power system. When thinking about this, it makes sense that risks abound on the delivery and the smoothing out of power to a computer system.

1. The building owners build office buildings to accommodate people, not machines. In Fig. 3.5 the power problems are summarized below. Owners tend to minimize services (for cost reasons) for a certain level of machines, telephones, etc. Most still use a design effort that will accommodate electric typewriters, adding machines, dictation equipment, etc. All of these items draw low power within the office environment. These owners overlook the power needs in today's office for PCs, Macintoshes, servers, laser printers, and other high-power consuming peripherals. Consequently, they undersize the utility that will lead to severe impact on the systems. If enough power is not available, the systems will malfunction through power dips, causing disk information retrieval to slow down, or causing the servers to crash. The proper power level is necessary to ensure equipment operates at optimal performance.

2. Power from the public utilities delivered through the building confines are prone to dips, sags, brownouts, blackouts, and spikes, as shown in Fig. 3.6. All computers are power sensitive. All of these conditions will or can cause them to act irregularly. If a user is in the midst of a major system backup and the power dips, the system could corrupt the data, rendering it totally unusable. Over the next decade the electrical needs of consumers will adversely be affected, with major brownouts and blackouts occurring. The electric companies are not building sufficient new generation capacities due to the following:

Problems with Building Power
1. Landlords and owners minimize utility space.
2. Occupied office space is maximized.
3. Tenants have PCs, Macintoshes, servers, and laser printers.
4. The office is built for standard electric office equipment:
■ Typewriters
■ Calculators
■ Dictation Equipment
Therefore more power is demanded than what is supplied, resulting in power dips. This causes major LAN problems.

Figure 3.5 Summary of the causes of many power problems.

Power Systems
Major Causes of LAN Downtime

- Sags
- Dips
- Spikes

Caution: Over the next ten years, severe shortages are predicted, due to lack of construction and increased consumer demand.

Figure 3.6 Power is a major contributor to LAN downtime.

- Cost

- Consumer pressure

- Confusion over what technology to use or build: fossil fuel or nuclear.

3. Other components added to this equation are major construction in and around the building and excessive consumption of power, causing disruptions or dips, sags, and brownouts.

4. Electric welding in the area can cause noise on the power line that will impact the LAN. The list can go on forever.

Figure 3.7 shows a disturbing fact. This graph represents the percentage of problems associated with different forms of power disturbances.

The protection of a power system can follow many different paths, depending on the money available and the critical nature of the LAN traffic. Some areas that can be developed to prevent outages are the following.

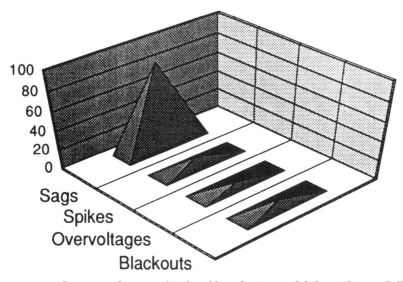

Figure 3.7 Summary of power related problems by type and failure. (*Source: Bell Labs and American Power Conversion, Inc.*)

Install uninterruptible power systems on every LAN server.   Servers, workstations, and printers are the critical components of the LAN. Protect them the best way possible. An uninterruptible power supply (UPS) can prevent many of the disruptions listed above. These systems are gradually becoming more reasonably priced and intelligent enough to work on a LAN. These UPS systems may take on various approaches, as follows.

1. *On-line UPS systems* cost more than other types of UPS but provide a power filtering and conditioning service. The on-line system sits in front of the LAN hardware all the time. A rectifier (a battery charger) converts the incoming commercial power that uses alternating current (AC) into direct current (DC). The converted DC power then charges a battery and feeds a device called an inverter. The inverter converts the DC power back to AC to run the attached LAN devices. Although this sounds complex, it is a proven technology and works well in the environment. The on-line UPS is shown in Fig. 3.8.

2. *A standby UPS* performs much the same activity as an on-line system but the significant difference is that it is not 100 percent uninterruptible. The standby system uses a relay switch on the output side. Under normal power conditions, power is provided to the LAN device from the commercial feed. Using a power monitor, the UPS only is engaged when the commercial feed is lost, activating the relay. The timing for this to happen is 2 to 10 ms, which allows for a very short outage potential. The power supply in your PC or server should have the capacity to store enough power inside it to accommodate this momentary outage, hopefully without any degradation of service. However, it really depends on the activity taking place within the system at the time. A 2- to 10-ms outage or dip occurring while a database index

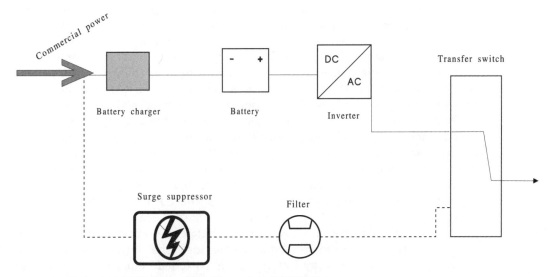

Figure 3.8   The on-line UPS system sits in front of the LAN equipment.

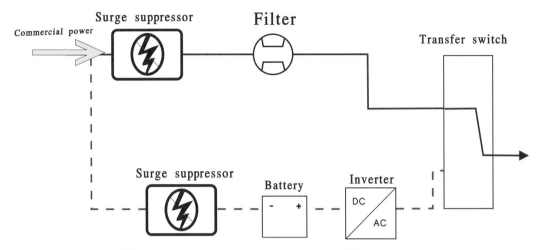

Figure 3.9 A standby UPS system kicks in when a commercial power failure occurs.

is being performed could potentially corrupt the index. No system is perfect; accept this as fact. Figure 3.9 is a representation of the standby UPS system. Always remember, regardless of the UPS used, all can fail, thereby eliminating the power backup system.

Surge protection and noise filters on the line can be provided separately but are often best served through the UPS. To get a better feel for this problem consider one of these scenarios.

When using a server on the network and an on-line UPS, the LAN receives all of its power from the UPS system. Any surges or impulses on the line can be filtered out of the environment before reaching the equipment. This is obviously the best approach satisfying most people's needs and concerns. But what if the UPS fails (and they do)? The power to the system will then revert to battery backup for some period of time before failing. In a LAN environment this is critical since the equipment can be spread around the organization and requires some monitoring service to alert the user and LAN administrator that something has happened. The timing for a UPS on a typical service will be from 15 to 60 minutes, based on the system and battery capacity involved. This is mostly a financial decision. What is needed is enough time to get the alarm, react to it, and provide for an orderly shutdown of the server or network. This means that all users should save their files and log out of their applications; all applications would then be closed and the server turned off.

Another scenario: instead of using the on-line UPS, a standby system is used. The UPS fails, but no interruption occurs because the commercial power still comes to the unit. However, what if the outage occurs and the power-monitoring equipment is malfunctioning? The battery would not be called for because the relay has not been switched. Thus the server goes down. This is merely a caution that even though a UPS is installed, no system is perfect. Never assume 100 percent protection.

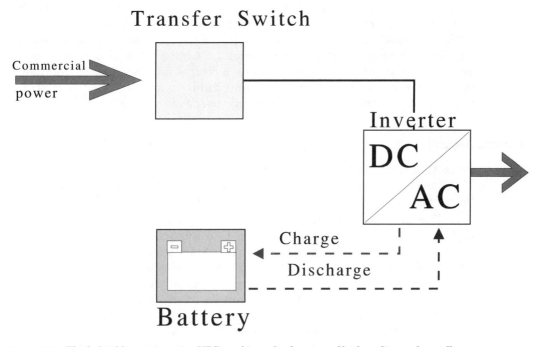

Figure 3.10 The hybrid line-interactive UPS combines the features of both on-line and standby systems.

3. *Hybrid systems* called line-interactive systems exist. These units allow the commercial power to feed the server, but the inverter is always on-line. This takes the characteristics of the two devices and melds them into a single function. This system line drawing is shown in Fig. 3.10.

The key issue here is to weigh the benefits and rules of both the UPS delivery and the cost of each. In the LAN arena, money is always a factor.

Should the power protection be provided at the individual server or other device, or through a centralized unit in an electrical closet? Our experience from the early data processing days shows that there is some benefit to providing very large units in a centralized room. However, the data processing equipment was also centralized. The LAN environment distributes the equipment and changes the rules for the need for power protection.

There are various ways to look at this; for example the comparison in Table 3.3 looks at the various sizes and costs for centralized and decentral-

TABLE 3.3   Comparison of a Single Unit vs Distributing Several Smaller Units

Unit	Size, kV · A	Number of units	Cost
Centralized	20	1	$15,000
Decentralized	0.4–0.6	No. of servers = 20	20 × $600 (av.) = $12,000

ized UPS systems. In this particular example, one would believe that the best solution is to distribute the UPS systems out to the servers.

The difference here is that the centralized unit has to be attached to the distributed internal electrical feeds around the building, whereas the smaller units are locally attached. Furthermore, a central unit failure will shut down the entire system while a local UPS unit failure may only affect part of the network. Which one best suits the specific needs of the LAN?

Hardware Security

Protecting the hardware covers many other areas that may not be obvious. This can include such areas as the physical security of the devices, prevention of loss via theft, the server environment, printers, and disk arrays.

1. The physical security within the environment most likely belongs under the domain of building security. Can intruders get into the environment undetected? Can they get by users unchallenged? Are work areas closed off from the general population? Do security services such as locked rooms, card access systems, etc., exist? If any of these areas are exposed, then speak to building and facilities management. Put strong policies and procedures in place to stop casual access to any equipment. Figure 3.11 is a collage of different security techniques that can be employed. By using security guards, sign-in sheets, and special locks on the doors to the server rooms, the LAN can be protected much better. However, reality sets in when financial implications are considered. Use whatever precautions that can be reasonably afforded.

Figure 3.11 Employ as much security as reasonably can be afforded in the LAN equipment rooms.

2. Theft is always a risk. The physical removal of a PC or server from the building should require an Act of Congress! Could someone walk into the building and walk out undetected or unchallenged? What if the thief wears a uniform, implying a maintenance or repair call? Do any avenues exist that would allow the stashing of equipment for later retrieval? If so, tighten up these procedures.

Does a system of passes for parcels exist for the removal of equipment from the office area? How well is it enforced? These areas are extreme, but think about cables, connectors, drives, and cards that could fit into a briefcase or large purse.

A network card removed from the computer room or from a PC could have consequential damage or long-term ramifications. Could a cleaner remove any equipment? Could equipment be lowered outside windows or stashed in trash dumpsters? Does security check this?

3. The servers need to be protected from theft and from casual access. Most LAN servers are PC based with keyboards attached. Could anyone walk up and access the server from the keyboard? Would employees question a stranger at a server? Lock these devices in a room. If access is needed, have a clearance or access procedure to ensure need and competency.

Once again the distribution of equipment throughout the building may prove to be a disadvantage over centralized control. Engage the following safeguards.

1. Keep logs of who can and cannot access the server room.

2. Keep logs of who goes in, when, and why.

3. Have an intrusion alarm on the entrances to server rooms.

4. Remove the keyboard to prevent unauthorized use, or have a lock out of the keyboard with password protection.

5. Put extra security and passwords on the servers.

6. Bolt server chassis to desk or use other similar techniques to secure them.

7. Double check all cable connectors. Make sure they are tight.

8. Check the cable system for taps, bridges, etc., that do not belong there.

9. Enforce a parcel pass system, check bags, boxes, briefcases, etc.

10. Have an escalation list for approval and checking of systems and components removal. A single person may not always be available.

Figure 3.12 can be used as a checklist of items to verify when checking hardware and closet security.

Printers

A new problem cropping up in the LAN arena is the physical loss of printers. As more people purchase inexpensive PCs (or acquire them in some

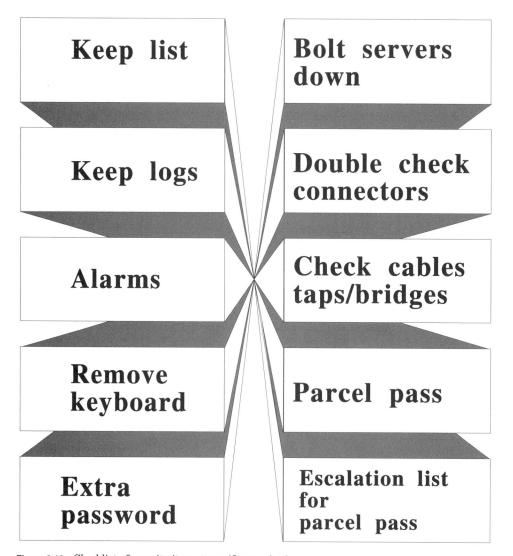

Figure 3.12 Checklist of security items to verify security in rooms.

other way) they realize that they have a need for printing. Laser printers (i.e., page printers) are becoming smaller, lighter, and more acceptable in the business community.

As these printers become less expensive, more of them are deployed within the office, thus becoming targets for theft. Secure them to the station they are attached to by bolting them to a desk or countertop.

Another concern with printers is the output of information. When confidential information must be printed, or when special paper is needed, the users need access to a workstation that is closer to the printers. No one wants to send confidential information to a printer located several hundred feet or a floor away. The fear is that the information could be viewed by unauthorized persons. Furthermore, when letterhead or other special paper is needed, the user will have to go to the printer and load the paper. Then the user runs back to his or her desk and sends the file to the printer, only to find that someone else has printed to the same device and used the special paper. This is annoying and wasteful. The frustrated user may abuse the printer for lack of something else to do about the problem.

Print queues are also a concern. If the queue is paused and work is stacking up, the user perceives the printer is broken. The user may then send the print job to another queue. This is fine for an immediate need, but the original job is still in queue. When restarted, the queue may have to be emptied or canceled. Can the individual user do this? Do they know how? Do you want them to do it? If the queue is restarted, what happens to the jobs in the queue? Figure 3.13 reveals the confidentiality problem with remote printing capabilities. As confidential material is sent to the printer, understand that anyone can pick it up and read it, or make a copy of it.

Disk Arrays or Disk Servers

These devices must also be protected from abuse, theft, and access. When the disks are readily accessible, a bump may cause a server to fail. Can the system be moved? Most likely not without risk. Therefore, these systems must be locked up in a protected environment with controlled access.

Other Areas of Concern

Can files be downloaded from the hard disk to a floppy or other hard drive on the network? Will this expose the LAN to failure or other breaches? In Fig. 3.14 the possibility of an employee or other person downloading files to a floppy disk introduces the risk of information being lost, possibly without being detected. This is a tougher problem to resolve since the access will be at a later date when this person is at his or her desk or outside the building, making it far more difficult to detect. Secure the data with the appropriate passwords and attributes for access and copying to prevent unwanted downloading. Keep these servers in a secured environment and lock the files from global or casual access.

Software Security

Beyond the physical protection of the rooms and the hardware comes the need to protect the software from abuse, misuse, and theft. This is a little more difficult to detect since copies can be made without being obvious. Issues concerning software protection are provided in Fig. 3.15 as a checklist and discussed in detail below.

1. Application software must be protected from crashes and from being bootlegged, violating copyright laws, etc. Enough material has been published

Figure 3.13 Confidential printing must be protected from unauthorized access and copying.

about the need to make working copies of the applications to prevent corruption and loss. The need also exists to protect the licensing and arbitrary copying of software which can create a problem of monumental impact. Can an application be loaded from any station on the network? Can virus protection be enforced? Do all users have floppy drives to load or unload software?

These issues can create major problems on the network side. If a corrupt application or an infected application is loaded, all users can suffer the conse-

Figure 3.14 Can files be downloaded to a floppy for later viewing and use?

quences. The best (but not necessarily easiest) way to protect the network is to use dumb terminals or diskless workstations. Any application that must be loaded can then be checked by the LAN administrator before loading. System configurations that may conflict with the software can be analyzed in advance before creating problems. Many PC applications could and do alter the config.sys, autoexec.bat files within systems. This altering could lock up or corrupt the network configuration, causing a disruption to the LAN. The reverse situation is that a LAN user copies an application from the LAN to a floppy for unauthorized or unlicensed use. This violates most copyright rules and could become a major issue.

2. Oftentimes users write to a disk that may have developed problems. At a later date the access and retrieval of their files becomes impossible; thus data integrity has been compromised. The LAN manager comes under fire because the data has been corrupted. In many cases this goes undetected when the write is done. A technique to read after write, similar to tape systems, can prevent some of these problems in the future. Utilities exist to recover some or all files that are corrupted or stored on inaccessible track sectors, etc. However, alerting messages must be used to detect this problem.

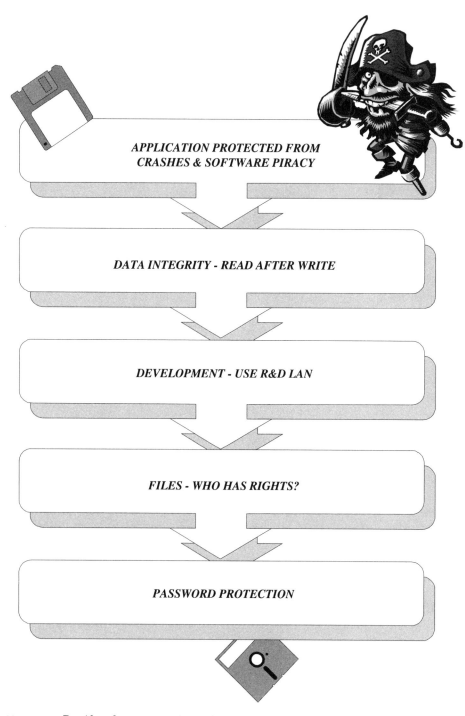

Figure 3.15 Provide software protection and security. Ban bootlegged copies of software from the LAN with strict enforcement procedures.

3. Any software developed in-house or outside should be checked prior to installation on a LAN. New applications should be *thoroughly* tested prior to certification. The LAN manager should have a research and development LAN available for the development and testing process; otherwise significant amounts of downtime could become the norm.

4. File security is an important issue and is connected with the three preceding items. How are users allowed to use the files on the LAN? Does the LAN manager give access to files? Or is this delegated to the owner of the file? If the owner grants rights on a network, does the owner know how to do so? Does the user know the differences between permission to:

- Read
- Write
- Edit
- Delete

Does a system exist that emulates a grandparenting effect? How many copies of a document or file are allowed to exist on the network? What rules are in effect to protect the files?

5. Passwords should be changed frequently by the user. A programmed arrangement based on 30 or 60 day intervals may suit the situation for some, whereas a 180 day arrangement may be more suitable for others. Note the following precautions.

- Change passwords regularly.
- Only allow a password combination once.
- Do not allow users to reuse the same password over and over.
- Put in double density on passwords.
- Protect confidential documents and applications with extra passwords.
- When a user's password expires, how is it enforced? at a particular time, upon access, or upon exit?
- If a password is breached or lost can you recover a file?
- If two users sign on with same log-on or password, does the system generate an alert or alarm message?

Threats

Threats to the LAN encompass many of the items listed above. Many of these are internal problems, but they *all* involve humans. The internal and external threats to the network can also include such things as shown in Fig. 3.16. The first three of these threats include the following:

1. *Viruses:* Everyone knows that this can wreak havoc on an individual station or an entire LAN. As long as applications or files can be loaded by users, the risk is high.

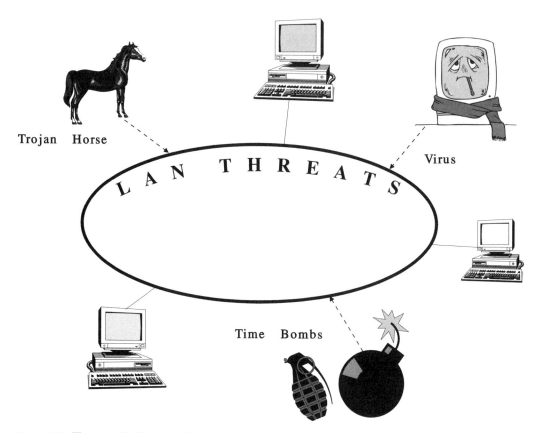

Trojan   Horse

Virus

Time   Bombs

Figure 3.16 Threat to LANs are major risks that are often overlooked. The first three threats discussed in text are included here.

2. *Time bombs:* Also called logic bombs, these are destructive codes that activate on a specified time or date. It seems that everyone is aware of the "Michelangelo" threat.[1] An attacker may penetrate the network and plant a bomb that can be set for the worst of times. This program can make it difficult to track or trap the attacker; it can be activated as a result of another event or it can throw confusion on a network allowing added time for some other destructive event to take place (i.e., a virus).

---

[1]"Michelangelo" was a time bomb that was set to "go off" or activate after a user logged onto a network or started a PC on March 13, 1992. Upon startup the bomb would then destroy the files and applications on the PC or network, causing significant damage and disruption to organizations. "Michelangelo" was to activate on the date of the famous artist's birthday. There are other time bombs that are programmed to be activated on specific dates (i.e., April 1, July 4, etc.) "Michelangelo" received a lot of press and produced hysteria in the industry. Therefore, the resultant damage was minimized. Many industry experts felt that the press overstated the risks and caused untold problems for PC owners and LAN administrators. But, who is to say what the results would have been without the awareness campaign?

3. *Trojan Horses:* As a Trojan Horse is activated, it is usually designed to take control of a user's privileges or access without the user suspecting it. Trojan Horses can transfer access privileges to another area or node in a network that is accessible by the person responsible. They can be used to destroy, delete data, or open secure files to others.

The next grouping of threats include the following four risks as shown in Fig. 3.17. Make no mistake, these are real threats.

4. *Browsing:* An intentional search for specific information (i.e., payroll, sales information, house numbers, client lists, etc.) that may be accessible. This can also be done on files that have been deleted but not yet overwritten. An example of this possibility cropped up on a popular bulletin board service (BBS). In order to make space for an application the disk may be rewritten with information already on it. The deleted files, still on the disk, become visible. By deleting the files, they were erased from the directory. But, since the disk had not been overwritten, the files were still there. So they were undeleted and visible to others who were restructuring the hard disk. Tools exist that allow this very simply. These inexpensive tools are readily available to a novice or a professional and pose significant risk.

5. *Salami attacks:* These threats occur in a covert manner. Rather than attacking user files or network access, this method is used to accumulate small pieces of information on a regular basis and store them in small quantities, thus preventing easy detection (hence the salami slice approach). In

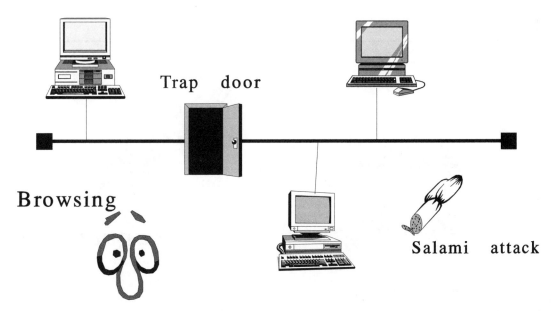

Figure 3.17 Additional threats to the LAN.

most cases a salami attack is conducted over longer periods of time and requires patience to discover.

6. *Hackers:* Always using random access to systems, these human threats can be both destructive and frustrating. Hackers are creative and intelligent, drawing their pleasure from defeating your systems and security. They can and will use any device available for their own gratification.

7. *Zappers:* These are people who write and plant macro codes or utility programs to bypass normal restart or recovery procedures.

The last grouping of threats is shown in Fig. 3.18. The issues include the following:

8. *Tailgaters:* These are attackers who may break in on remote access service. When a user is remotely accessing your network the communications may be abruptly disconnected. The communications controller then reconnects the original session to the next user on the queue, called a tailgater.

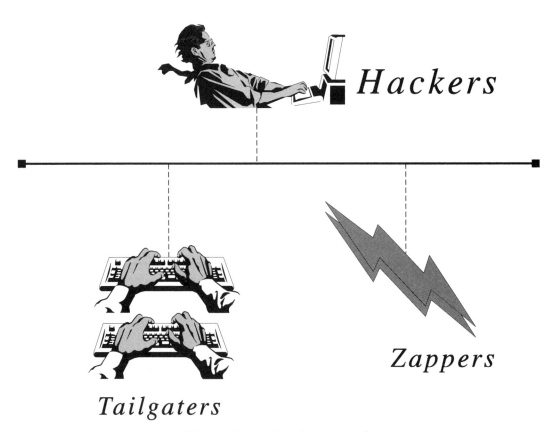

Figure 3.18   Added threats to the LAN; do not limit risks to these issues alone.

9. *Trapdoors:* These techniques are surprisingly still quite common. Access procedures and security measures are bypassed so that developers and server personnel can gain easy and instant access to a network service bypassing normal security and log-on procedures. The movie *War Games* popularized this technique.

Chapter

4

Connectivity Issues

In any local area network, users ultimately mature in their desires and understanding of how the network functions. Thus, evolution states that they all face the same needs. Whether users in the local environment have reached that point yet is only a matter of time. It is not a question of *what* they want but *when* they will want it. Users will want the ability to connect to a server or an application on another LAN or segment, etc. That other LAN may be in the same building or 3,000 miles away. Regardless of the distance, that connectivity will have to be met. If the LAN administrator does not provide the connectivity solution, users will find their own solutions to their perceived or real needs.

However, once these devices (files, applications, etc.) are connected, the LAN administrator's added responsibilities begin to magnify. No longer will the LAN disaster recovery plan and protection process be localized to a single network or building. The planning phases will include access to other locations that may or may not be under their control and must be considered. These needs will involve the required issues minimally as follows:

LAN to LAN	Simply stated, a bridging or routing capability between two LANs. Users may find an occasional need to access a remote server for a specific application. Usually this is an in-house requirement that can be easily met but may eventually require heavy access to a server or user group on a regular basis.
LAN to WAN (wide area network)	Linking LANs across a much wider area network through bridges, routers, or synchronous communications links. These links can be lease lines, dial-up lines, or access to packet-switching or frame relay services on a value-added network. In this scenario the LAN will be exposed to internal or external net-

work glitches that can disrupt the orderly flow of LAN traffic.

Public switched access

Asynchronous communications links across a telephone company network. Such connections could expose the LAN to hackers, vandals, or other disruptions on an inbound or outbound basis.

Interoperabilities on the network through gateway functions or router functions

Gateways allow access between disparate networks while routers switch traffic between LANs via dial-up lines, leased lines, or the more traditional X.25 services; newer capabilities include frame relay.

Figure 4.1 is a generic representation of these services on a network platform. This includes the four basic areas where this connectivity can be expos-

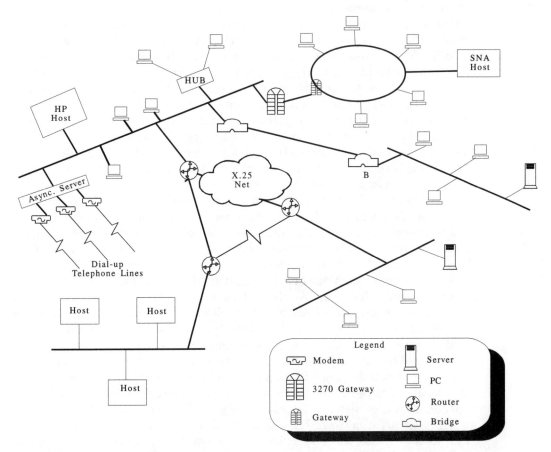

Figure 4.1   Connectivity issues involve bridges, routers, gateways, or asynchronous switched access.

ed. As more access is given, security protection systems and other risks increase exponentially.

LAN to LAN Connectivity

Within a department or a building the issue of connectivity is fairly straight-forward. Depending on the responsible person chartered with the administration of the LAN, the connectivity can be as easy or as complex as the situation dictates. However, the LAN manager must be proactive in providing the service. If not, the users will source out their own solutions. It is not uncommon for users to go outside the organization to find components necessary to connect their LANs. Vendors of systems or system integrators have been retained to provide a solution, bypassing the LAN manager. These items are part of the users' budget or expensed on a different budget. This situation tends to pose many other risks due to compatibility issues, lack of knowledge to maintain the system by the user, or general degradation of the network due to excessive traffic.

Suppose two operating departments wish to share files and applications between them. They may also desire an electronic mail service between users on both LANs. Extending this scenario a step further and assuming that the LAN manager or administrator is in the MIS department, MIS determines that the groups are on similar operating systems and using the same topologies. Thus, the connection can be simplified through the use of a bridge. The bridge can be added between the two LANs and can provide the necessary connectivity. Figure 4.2 is a representation of this sharing of files and applications across a bridging mechanism. Users go through a handshake to share their files.

But what if the network fails? How dependent on the use of file swapping and sharing have the two departments become? As with any service the initial installation is for a casual access or simplified operation. However, after the service is available, the users become more dependent on the service. This is difficult to assess unless an audit trail of usage statistics can be generated. Is mission critical information flowing between the two LANs? If so, the protection and recovery will be escalated to much higher levels. Most networks start out as administrative. However, they evolve into devices to run the core business functionality of the department. This complicates the planning process, unless the LAN manager clearly understands what services have migrated onto the LAN. Additionally, the users may have bought or developed proprietary or custom applications that are not documented well and that will be extremely difficult to recover in the event of a disaster.

Therefore when dealing with this LAN to LAN connection the movement of information between the LANs should be watched closely. Frequently, departmental needs change; consequently, traffic needs also change. Many an administrative network has experienced severe bottlenecks due to shifts in needs and access. More specifically, once two LANs are connected, if all of the write/read activity is to a server on the other LAN, congestion can occur at the access point (e.g., the bridge). Any network problems that crop up may

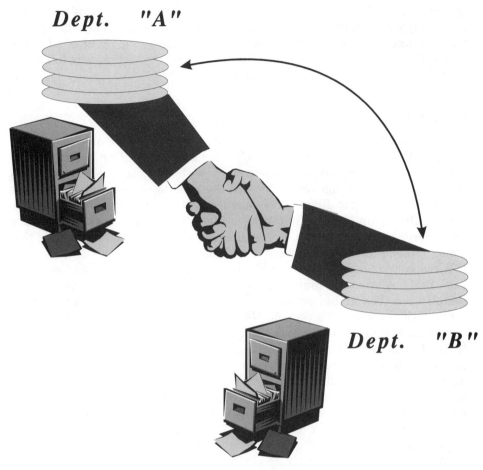

Figure 4.2 As departments need to share files, connecting LANs becomes necessary.

be the result of this access or congestion. This creates the need to move the users to the other LAN.

Before going any further, it must be stated that an assumption has been made that the users on both LANs are in the same building (or local area) *and* the same company. If, however, the two departments are from different companies, the entire connectivity and recovery planning picture changes dramatically. Administrators of both LANs must come to mutual agreement on the provision, protection, and recovery processes after the event of a network failure. A coordinated set of plans must be drawn up to handle efforts that must be applied so that the teams on both ends agree to the priority levels, the critical nature of the information, and the resources necessary to support this connection. This can be done by a handshake arrangement *or* by a well-documented letter of agreement between the two organizations. The latter is the preferred choice for protection against future misunderstanding

and priority shifts based on the organization's needs. The best rule of protocol is to get it in writing. Be prepared to accept full responsibility to recover
the LAN and associated connections and equipment in the event of a failure.
If the other organization shifts priorities, the LAN connection may fall below
the priority requirements of your organization.

LAN to WAN Connectivity

Local area to wide(r) area connectivity may take on a whole new perspective
in protecting and recovering the LAN after a disaster. First, when dealing
with the same company and department a physical link can be installed
under specific *direct* control. Even with the need to connect among multiple
LANs in a mix of buildings and companies, the control process can be handled relatively simply. This of course assumes that a logical thought process
has been applied in advance. Assuming that the LAN falls under this
domain, the necessary interconnectivity rests there too.

When connecting through a wide area network the component selection
(e.g., bridges, brouters, routers, gateways, and modem pools) becomes more
complicated. The ability exists to add robustness through the appropriate
connectivity component selection. Never assume that someone else will take
responsibility for recovering the network. This assumption could lead to
future finger-pointing and confusion at a time when it is least needed. The
primary role is to get back into business as soon as possible. However, a new
cast of players is introduced (local-exchange carriers, interexchange carriers,
and/or private value-added suppliers). These carriers are concerned with the
recovery process but have other priorities to deal with. Individual LANs and
WANs will normally be a low priority. This is especially true if a major or
regional disaster strikes. Figure 4.3 is an illustration of a wide area network
connecting local area networks.

Significantly more exposure points exist through the use of a telephone
company's and long distance company's network facilities to connect LANs
together across a wide area network. Whether the choice is to lease lines
(consisting of copper, fiber, or radio systems) or to use the network supplier's
network for circuit-switched connections such as dial-up telephone lines,
packet switching, frame relay, etc., the multiple points of failure that could
impact the network has exponentially increased. The planning and recovery
process will now involve at *least* three (if not more) different companies. Each
of these will have a separate set of recovery priorities. Control shifts to these
new inductees during the planning process to recover network facilities in
the event of a major network failure. Questioning these carriers on their ability to recover the network could serve longer-term needs. Gaining commitment from them in advance will better serve long-term needs.

Public Network Access

This is another option available to help solve the connectivity problem. It may
also prove to be a value-added service if or when a disaster eventually strikes.

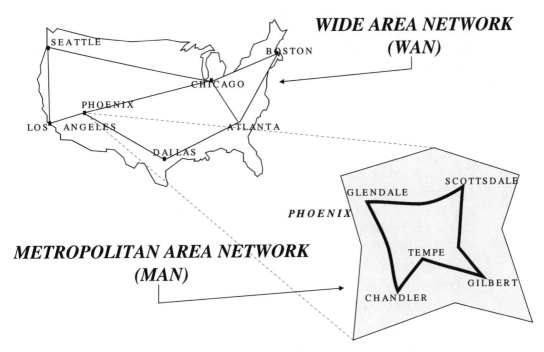

Figure 4.3 LAN to WAN connectivity introduces new complexities in preparing for disasters.

Those leased lines or other services (regardless of technology) *can* and *will* fail. Depending on the disaster the network may be adversely affected through the incident. As a result optional and dynamic pipes (or lines) must be made available on a dial-up basis to overcome individual crisis situations. It would only be through a major disaster (as related in the June 26 and September 17, 1991, and other disasters[1]) that dial-up services could be significantly impacted. Dial-up network or the leased-line services from the net-

---

[1]On June 26, 1991, the Bell System (specifically Bell Atlantic) had a major outage on their network. The problem came from a "bug" in the Signaling System 7 network (SS7) where no calls could be processed for 10 hours in a four-state area (MD, DC, VA, WVA). The software bug caused the systems to lose their reference pointers to place calls from customer locations through the Bell system.

On that same day in Los Angeles, a similar problem with a bug caused 50 central offices to fail in processing calls. Therefore all local deal-up calls were unable to get through to the called party.

On September 17, 1991, AT&T had a major power failure in their regional switching center just outside New York City. The power company put them on backup power (AT&T's own generators) but the system failed. Consequently AT&T's office failed for 8 hours due to lack of power. The air traffic control system for Kennedy, La Guardia, and Newark airports were all affected; thus planes could not be guided in to land. With this, long distance calls placed from this area could not be processed either. This caused major disruptions for customers getting their data through from one network to another.

The telephone network is just a series of computers linked together with lines. When it fails, the rest of the consumer and business users also suffer.

work suppliers have accounted for over 114 network failures over a 12-month period, which should cause some concern to a LAN manager. Regardless of the cause, whether from the newer signaling systems, cable cuts, natural disasters, or human error, the primary point is that the leased-line and dial-up networks are going to be exposed to outages and be influenced and controlled from the outside. Recovery and restoral procedures will have to include alternatives to these potential risks. These alternatives can include:

1. Multiple carriers, possibly on the long distance portion of the connections.

2. Diverse routing through the central office via alternate physical routes through either the purchase of rights-of-way or through a long-term lease arrangement.

3. Different technologies, land-line, satellite, or radio systems, integrated into a network scheme. These should minimize most of the risks associated with failures.

4. Shared facilities with others.

5. Different techniques, such as dial backup for lease lines, frame relay, and packet switching, through the various carriers.

Opportunities exist in many forms, but the plan must address these *in advance*. The worst time to consider alternatives is after a disaster strikes. Remember, the planning process must highlight all risks and orchestrate the reconnection process. In Fig. 4.4 the alternative is highlighted as a means of multiple carriers protecting WAN capacities. However, since this is only a portion of the risks that threaten network connectivity, the use of diverse routing is shown in Fig. 4.5. To overcome the risk of the local connection or the long-haul portion of the network, alternative technologies are shown in Fig. 4.6. Lastly, fallback capabilities such as X.25 packet-switching techniques through the use of routers is shown in Fig. 4.7.

No one solution will meet all needs. The LAN manager must determine the risks of each attachment and plan the appropriate fallback or recovery process. However, care must be taken to minimize expense while maximizing use of time. Many of the carriers do offer protection services on their networks, but they must be coordinated in advance.

Communications Equipment

Interconnecting LANs, as already stated, is an inevitable evolution of the network. No matter how much control and containment exists, users will want to connect to other services whether through the techniques noted above or any other methods. The LAN manager must carefully select the technologies. As the selection is considered, care must be taken to minimize the risks. As the need to connect various entities arises, the available options are as follows.

1. LAN to LAN

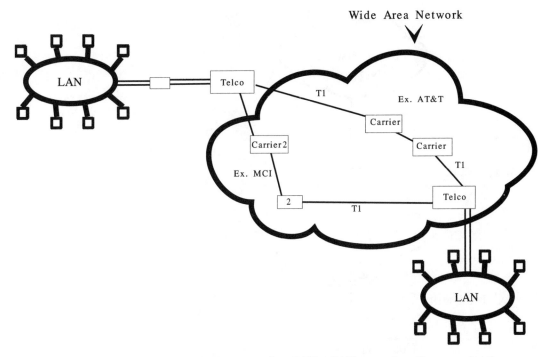

Figure 4.4   Alternatives exist to use multiple carriers from LAN to LAN across the wide area network.

2. LAN to MAN [Metropolitan area network: LANs connected in a statistical metropolitan serving area (SMSA)]

3. LAN to WAN

To provide this form of connectivity, various LAN communications components exist. The decision to use any or all of these pieces really depends on the individual network and the types of services required. In every situation, the mix of services is contingent upon the LAN operating characteristics and the need for transparency. The components are as follows:

1. Bridges

2. Routers

3. Gateways

4. Brouters (bridging routers)

5. Asynchronous communications servers

To protect both the connectivity and security during the LAN communications process, a true understanding must go into the selection process. Each has variable risks and recovery procedures that must be carefully considered. Let us look at these one at a time.

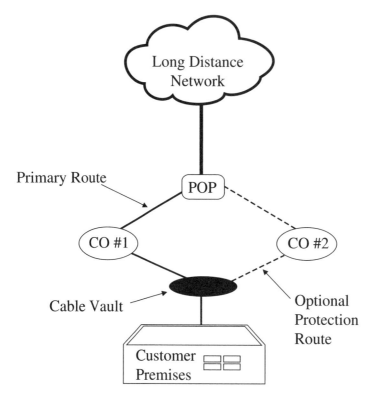

Figure 4.5 Diverse routing from the building to the local central office. These can be through lease facilities or rights-of-way access. Checking carrier maps will help determine if feeds from two COs will help prevent a major disruption.

Bridges

Bridges are the simplest form of connecting LANs because they deal at the lower layers of the OSI model (at the physical and link layers), which is a model designed to provide transparent communications in the environment. The International Standards Organization (ISO) developed a model to allow for the transparent transfer of data from device to device. This model called the Open Systems Interconnect reference (OSI) is a seven layered architecture. If all manufacturers abide by the model on a layer-by-layer basis, the transfer of information will be transparent. Since only the lower levels of this model are involved, it would be safe to assume that a physical connection is required between two similar and compatible LANs. Regardless of distance, the bridge will act as a filtering and forwarding agent, shipping only packets that must be forwarded across a network boundary. Any packets that do not need to cross the boundary (stay local) will be filtered by the bridge. Some bridges are merely two LAN cards in a single PC that spew every data packet between both LAN segments. This could cause heavy traffic on the LAN

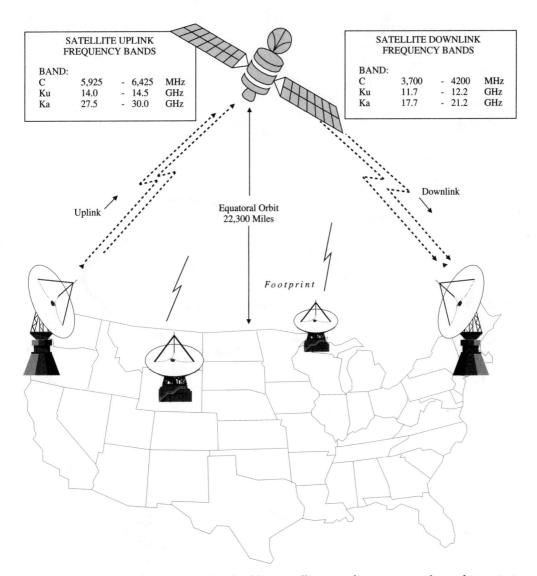

Figure 4.6 Alternative technologies such as land-line, satellite, or radio systems can be used to protect the network.

and subsequent disruptions to network traffic. LAN managers should be wary of this bridge characteristic.

If two Ethernets are connected between the east coast and the southwest via a bridge, certain assumptions can be made: see Fig. 4.8 as a summary of the following information.

1. Since both LANs are running Ethernet, the packet types are the same. Ethernet packets of the same size, type, and content are assumed here.

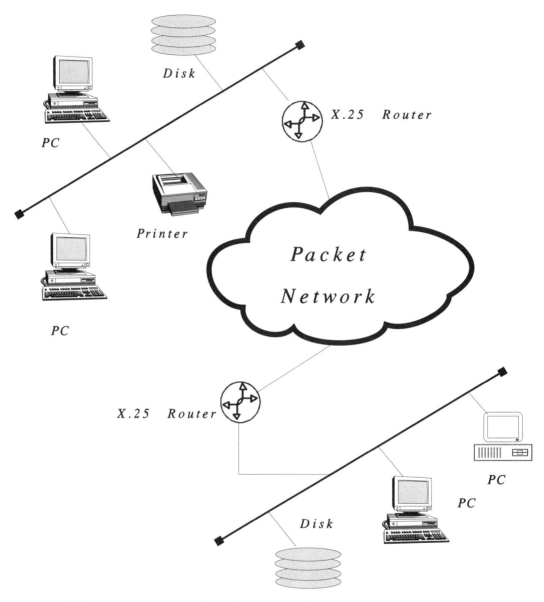

Figure 4.7   X.25 or frame-relay services through routers can offer network recovery facilities in the event of a facility failure.

Standards exist that make this statement true. On top of the Ethernet, a Novell operating system is installed.

2. The link between the LANs will most likely be a dedicated line operating at 56 or 64 kbits/s or T1 (1.544 Mbits/s). The distance is merely added to run an outside link. It could also be a simple connection on the same floor in an office building. For now, however, the distant ends will be used.

Checklist for Using a Bridge for Link Recovery
1. What topology does the LAN work with? ❑ Ethernet ❑ Token ring ❑ Arcnet
2. Type of lines used to link LANs: ❑ 56 kbits/s ❑ 64 kbits/s ❑ T1 @ 1.544 Mbits/s ❑ Other
3. Do bridges use the same protocol? ❑ Yes   ❑ No
4. Do bridges use a learning or spanning tree protocol? ❑ Yes   ❑ No
5. Is the bridge used to filter and forward? ❑ Yes   ❑ No
6. (a) Can the bridge balance its load during heavy traffic conditions? ❑ Yes   ❑ No (b) If multilinks are available, will the bridge use alternate routes? ❑ Yes   ❑ No Does the bridge build spanning tree database? ❑ Yes   ❑ No How many hops will the bridge handle? _____
7. If the link fails, does the bridge have intelligence or line-sensing equipment? ❑ Yes   ❑ No
8. Will the bridge send out an alarm to a console or to a pager? ❑ Yes   ❑ No
9. (a) Can the bridge prioritize packets on the network if congestion or link failure occurs? ❑ Yes   ❑ No (b) Does this happen on a node-by-node basis or server basis? ❑ Node by node   ❑ Server
10. If power failure, what memory does the bridge use? ❑ Nonvolatile   ❑ Floppy   ❑ Tape   ❑ Hard Disk
11. After a link or power outage is restored, does the bridge restart from memory automatically? ❑ Yes   ❑ No

Figure 4.8 Link recovery between bridges.

12. Are buffers available to log errors, collisions, lost packets, or other usage statistics? ❏ Yes   ❏ No
13. If a remote access modem (dial-up) is available, can a terminal be used to check statistics, reconfigure, or reset or restart? ❏ Yes   ❏ No
14. Will automatic reload or synchronization take place if the system is moved? ❏ Yes   ❏ No How long will it take? _____
15. If a link fails and a lower-speed connection is available, will the bridge readily adapt to the new speed? ❏ Yes   ❏ No
16. Are multiple levels of passwords available for dial access from remote? ❏ Yes   ❏ No
17. Will the bridge shut down automatically if two or three remote attempts are made? ❏ Yes   ❏ No

Figure 4.8 (*Continued*)   Link recovery between bridges.

3. The bridges on both ends use the same protocols and are manufactured by the same vendor. A LAN manager needs to keep similar products so that other problems such as incompatible equipment will not cause adverse effects on the LAN.

4. Self-learning bridges with a spanning tree algorithm are used. This does not specify a requirement but will aid in network administration and control. The spanning tree protocol aids the bridge in recognizing the distant ends and devices attached to the LANs.

5. In order for the bridge to filter and forward packets, it must build a database of all nodes on the network. As a packet is sent out or a message address is received, the bridge will "learn" where the end devices are and remember them. A table of addresses is kept in memory. Nonvolatile memory is preferred.

6. In the event of congestion on the network, the bridge can use a load-balancing technique to shed load during heavy traffic periods. If multiple links exist, the bridge can use these links to share the load.

7. In the event a link goes down, can the bridge forward traffic to another link (such as a leased-line or an automatic dial backup circuit via a modem)?

8. Ensure that the bridge will send out an alarm if circuit failure occurs. The alarm could be to a local or remote console. Even better, look for a system that can send out a pager message.

9. If heavy congestion exists on the network, the bridge should be intelligent enough to handle the prioritization of packets. This could be done on a node-by-node or server basis. The excess traffic that is noncritical will be discarded.

10. In case of a power outage, the bridge will use nonvolatile memory. A tape, floppy, or hard disk should be used to store the network configuration information.

11. After a link or power outage is restored, the bridge should automatically restart from its stored load without the need for human intervention.

12. Buffers will be provided to log errors, collisions, lost packets, and usage statistics for later review.

13. The bridge may be accessed via the leased line or a modem (dial-up) from a remote site to obtain the traffic statistics, to provide reconfiguration, or to clear the buffers and restart the system.

14. In the event the bridge must be moved to a new site, the timing to recreate the database will be minimal. Therefore it can be ready to use after minor timing, synchronization, and restarts.

15. In the event a link speed (56 to 64 kbits/s, T1, etc.) is higher than the fail-over speed (9.6 to 56 kbits/s dial-ups), the system must adjust to the new speed as appropriate. If it does not, the fail-over link must be of similar speed. This could prove costly and could jeopardize the installation of redundant links. The LAN manager must verify this operation before relying on such a service.

16. If a dial access from a remote location exists, multiple levels of passwords can be used before gaining access to the bridge. Remember that if dial access is available, others could also gain access. A disastrous situation could result if hackers gain entrance to the LAN by the very system put there to protect it.

17. If repeated attempts (two, three, or four) are made and fail, the bridge will shut down the access automatically. Never lose sight of the devious minds of others. Vandals and hackers love a challenge! They will spend significant time and make many attempts to usurp any security system they encounter. By using a "kill" program to turn off a modem after three consecutive password failures, the system can be better protected. Still better would be an automatic dial-out to a pager on the console if the kill system is activated. This alerts the LAN manager of penetration attempts immediately.

Routers

Routers are more sophisticated and complex to manage than a bridge. Since the router is a higher-level communications device than the bridge, the router works at the bottom three layers of the OSI model, one layer higher than the bridge. The network layer is introduced here, which allows the router to send packets across multiple routes or paths between two points. Some of these routes may go through intermediate nodes (or routers) in the LAN connection.

Routers are involved in sending the packet to the distant end and ensuring data integrity and recovery. They are responsible for setting up a database of all links (nodes) they can communicate with. Sometimes the router will only be concerned with sending the packet to its downstream neighbor (router), which will know how to continue the communications process or the flow of packets. When building a database, the router can send out a "discovery packet" to learn the addresses of nodes at remote ends. This overcomes the bridge's need to wait for a packet to be sent on the network and delivered before learning of the device or user.

The router can use several routes to get the packet (as many that exist) regardless of a link failure, provided that all links into a specific LAN do not fail at the same time. This, however, could happen with a cable cut into the building or a building disaster serving the router access. Care must be taken to prevent such interruption of service as much as possible.

Routers are by far more sophisticated than basic bridges. A bridge learns just what other end it is attached to and it must be physically attached. A router does not have to be physically attached and can select any route that is available to it. Since this technique does not depend on a physical connection, it is more robust in its delivery of packets after a link failure. The discovery packet is used to determine how packets might get from node to node. In Fig. 4.9 a series of routes are built in the router's database. In Fig. 4.10 a database is built showing how node 7 on LAN 1 can connect to node 20 on LAN 5. A fictitious cost is assigned as a value of connecting the LANs together and the router is set to choose the least-cost route. However, if a

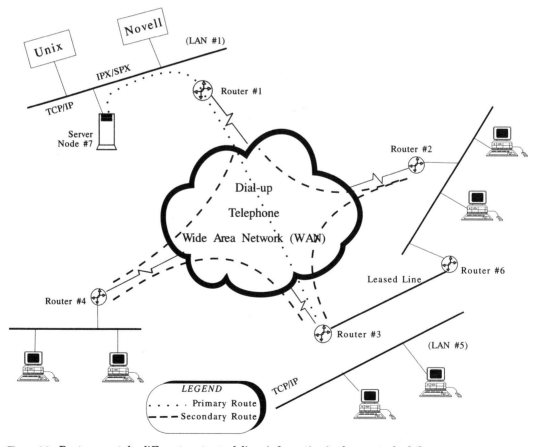

Figure 4.9 Routers can take different routes to deliver information in the event of a failure.

LAN 1 to LAN 5	$ Cost
1 to 3	1.00
1 to 2 to 3	3.00
1 to 4 to 3	5.00
1 to 6 to 3	2.00

Figure 4.10 A database of connections from LAN to LAN is created using a fictitious cost to prioritize circuit choices.

failure occurs, the router's first choice will be from 1 to 3 with a secondary route from 1 to 6 to 3 based on a cost of $2. This means that the secondary route to get from LAN to LAN will readily be available. However, to maintain the database and be aware of the availability of links, the routers constantly generate idle "chitchat" by sending out packets to each other. This idle chatter on the network could cause a significant amount of congestion on the network.

An alternative to this risk is to use two possible routes to the network. The first set of connections shown in Fig. 4.9 were over telephone company facilities. In Fig. 4.11 a second choice would be to route to an X.25 packet-switching network. A router can handle multiple protocols; therefore the system can carry Ethernet packets to an X.25 network quickly.

Gateways

Gateways offer the ability to provide LAN connectivity between different networks. From an OSI perspective the gateway covers the full seven-layer spectrum all the way to the application layer. A summary of these three techniques is shown in Fig. 4.12.

The gateway can provide access from a ring, a bus, a network running Novell, or another system to an X.25, SNA, or another system. Since this works at the application layer the gateway can place a 3274 controller as a node on a ring, or an IBM host system can be integrated with the LAN using a SNA gateway service.

A critical application running on a host can be protected well since host applications have been developed over the years. A dual attached node (PC) can be used with a token ring card on a LAN accessing the host as shown in Fig. 4.13. The secondary connection would be a direct cable to the 3X74 cluster controller.

In the event of a cable cut on the coaxial 3270 run, a node can use the LAN, pass through the gateway, and address the critical application. An

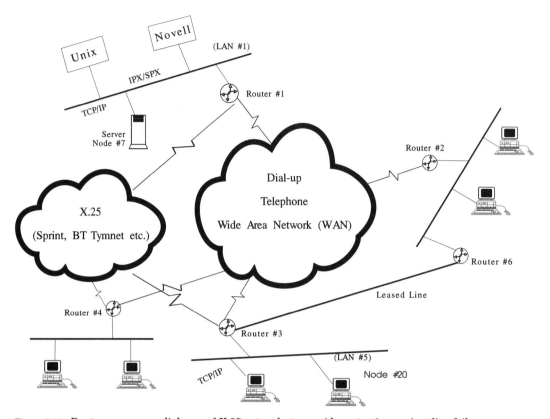

Figure 4.11 Routers can access dial-up and X.25 networks to provide protection against line failure.

Ethernet can also be used by connecting a 3270 gateway on the Ethernet to a front-end processor. This is shown in Fig. 4.14 with an SDLC link used from the LAN to the front-end processor.

Once again, if the link (coaxial cable) between the PC and the cluster controller or between the cluster controller and the front-end processor fails, access can be gained through the gateway.

Once the decision is made to provide mission critical or noncritical services on the LAN via gateway access, the LAN itself becomes mission critical. Consequently the gateway must be protected to prevent disruption to service.

Still another gateway service is the X.25 capability to access the LAN from remote sites. Using a carrier (Tymnet, Telenet, etc.) the entire LAN can become bottled up due to a network or gateway failure. This can disrupt more than the LAN primary users.

The above scenarios show these gateway functions on a more localized basis. However, if the access is through a remote node on a LAN, equal priority must be addressed. A PC residing on a LAN in a distant city can access the host application through the remote gateway over a leased line. If, however, the leased line fails, the gateway might use the X.25 or a frame-relay

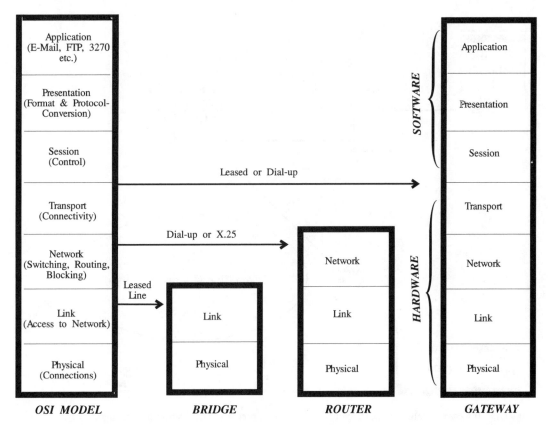

Figure 4.12 Bridges operate at the lower levels of OSI, while a router operates at the higher levels; gateways cover the full-layer model.

technique to recover quickly. This means that the gateway must support multiple conversions and protocols. In this particular instance it required X.25, frame-relay, and SNA capabilities. Figure 4.15 is a representation of these protocols on the networks through two different gateway functions.

In Figure 4.16, a multiprotocol gateway with X.25 output can carry SDLC packets to the host and TCP/IP or IPX/SPX packets back to the terminal device. (TCP/IP is a set of protocols used in many worlds, such as UNIX, OS/2 LANs, Novell LANs, etc. IPX/SPX is a proprietary set of protocols used by Novell in implementing their LAN operating systems.) This will prevent major disruptions on the network without shutting down.

Modem Communications

When the physical side of the LAN communications is being reviewed, modems should be added in the items being checked. One of the most overlooked technologies and possibly the most misunderstood is the ability to use modems for access onto and off of the network. These devices cover the gambit, since they

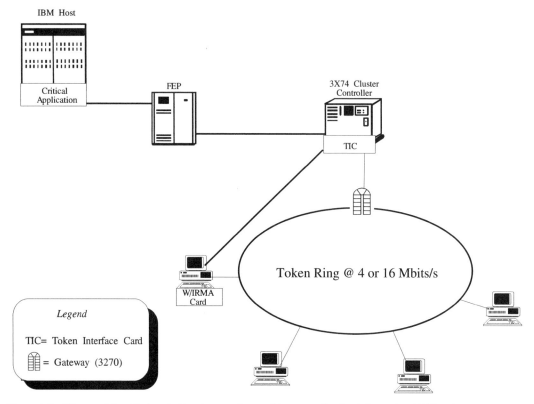

Figure 4.13 If a coaxial cable is cut, the ring can be used to access the host through a 3270 gateway.

can be useful as a recovery tool yet can also be a point of penetration into the network from unwanted users. Thus, a tool to provide backup recovery for communications after a disaster may well become the vehicle to create a disaster.

As physical devices connected to the LAN, modems can act as conduits for electric spikes on a telephone line. They also act as a doorway to the LAN through which, unfortunately, an experienced hacker can gain easy access. As a matter of course, a hacker can use a random or sequential dial access program and compile a list of all types of modems on lines within a range. Figure 4.17 is a summary of the sequence used by a hacker. In case any doubt exists, a good hacker can usually get through normal security systems in less than one hour.

All communications services must be protected against multiple threats. These include the ability to provide the following:

1. Lightning protection from coming in through the line and destroying your equipment. A surge bar as shown in Fig. 4.18 can protect against these spikes. Modems have limited surge and lightning protection. Installing as much more protection as possible goes a long way.

2. Prevention of dial access to hackers who can penetrate your network. This includes security protection systems such as changing authentication key systems.

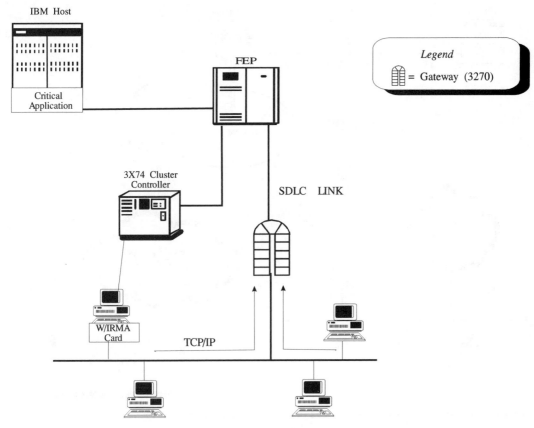

Figure 4.14 Access to the host from an Ethernet can be provided through a gateway.

3. Prevention of dial access to bulletin board services whereby a virus can be downline-loaded onto the LAN. This means that the service must be restricted or the information being downloaded must be sent to a "quarantine area" on a separate disk until it can be checked out for viruses.

4. Security against unwarranted access from disgruntled employees who can trash files. This requires immediate password and modem number changes. Preferably this will be done before an employee leaves and has a chance to access the network.

Modem communications through pools or asynchronous gateway connections are discussed in the following sections.

Dial-in access

Dial-in access to the network must be protected by all means available. This is the single largest exposure point on the LAN. Users want access to the network from a remote site and wish around-the-clock availability with little or no protection from penetration. However, the more they demand uninhibited access, the greater the risks facing the LAN adminis-

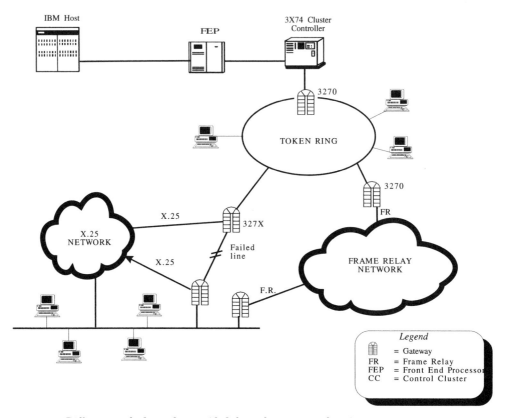

Figure 4.15   Different methods can be provided through a gateway function.

trator. First and foremost, the LAN is a compilation of network services and files that are private to the organization. The arbitrary access by any user (or hacker) exposes the organization to high risk. A disaster can be described as a loss of security or a loss of information. Either of these are risks based on dial-in modems. First, a hacker may dial in and use the attack methods mentioned earlier (browsing, salami, etc.). Being undetected, hackers can then rummage around the network at will. Second, a virus-infected file or floppy can be communicated up to the file server from a remote user through dial-in access. Third, a disgruntled employee can deliberately access the network and wreak havoc on the network with files, servers, or other services.

Therefore the following steps must be taken at a minimum to protect the network and files (these actions are summarized in Fig. 4.19 as a checklist):

1. Identify all users who have access to the LAN from off-site and compile a database of who they are, their numbers, and the services they are allowed to access. Check to ensure that multiple log-ons are not used simultaneously. See Fig. 4.20.

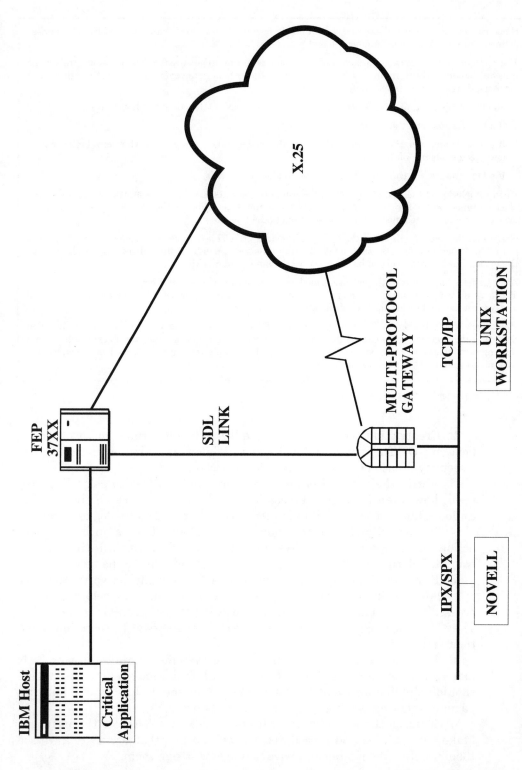

Figure 4.16 Multiprotocol gateways can prevent major disruptions on the network.

1. Hackers use inexpensive electronics and a single computer program written in BASIC. The hacker defines a range of telephone numbers to dial [e.g., (415) 201-1000–9999].

2. The computer will then rapid-dial each of the numbers in the range specified. Using a single telephone line and modem, the computer will select one line at a time. However, using multiple lines and modems, more can be handled simultaneously.

3. As the line rings the computer is poised and waits for an answer. Depending on the answer:

   a. If a human answers, the computer will hang up.

   b. If a modem answers, the computer listens for the modem tones, recognizing whether a high-speed or low-speed modem is on the line.

   c. If a fax answers, the computer hangs up, but makes a note of it.

4. Upon completion of the dialing sequence, the computer outputs to a screen or a printer a complete list of each number in the range. Furthermore, the list contains notations on the type of device (i.e., human, low-speed or high-speed modems, or fax) attached to the line.

5. Generally, the hacker will select the low-speed devices first. This is usually a sign of a small to moderate LAN. Experience dictates that small- or medium-sized LANs have limited security in place. Thus the hacker has a target and can be on the network simply and quickly.

6. After penetrating LANs with low-speed modems, the hacker shifts attention to high-speed modems. These typify user networks with more access and hopefully more security. Unfortunately, this is not always true, so the hacker has little difficulty gaining access.

7. However, some networks do have better security. These pose a challenge to the hacker's prowess in defeating the security. This can be disastrous to an end-user network, since the hacker will leave a "calling card" to prove how good he or she is. This can be in the form of a virus, time bomb, or a blatant mail message left on the administrator's screen.

Figure 4.17   Sequence of events used by hackers.

2. Create audit trails of any remote log-on by date, time, and user identification (ID). This audit trail will assist in follow-up activities in the event of a security breach or if any foul play is suspected.

3. Verify the usage. Contact the users and feel free to ask what they need, how often they need access, and if the accesses are all theirs. This could highlight a problem if users deny accesses based on the log-on activity.

4. Periodically change the telephone numbers. This is an inexpensive action, but if the numbers have made it on a bulletin board, time can be bought. Ask the telephone company for numbers in other exchanges.

5. If automatic number identification (ANI) is available, verify the user by a database inquiry, that is, capture the caller's number before granting access. This is an additional security procedure that is invisible to the caller. They therefore can eliminate one extra set of password protection, even though it exists.

6. Provide a system that will alert the systems administrator or LAN manager or any prolonged use of the network by a remote site or user. This should establish an event log for tracking and should have a dial-out to a pager or other messaging system during off hours.

7. Enforce the security procedures necessary to protect the network. Make sure policies and procedures are written and that users are aware of them. Publish them often and have management sign them.

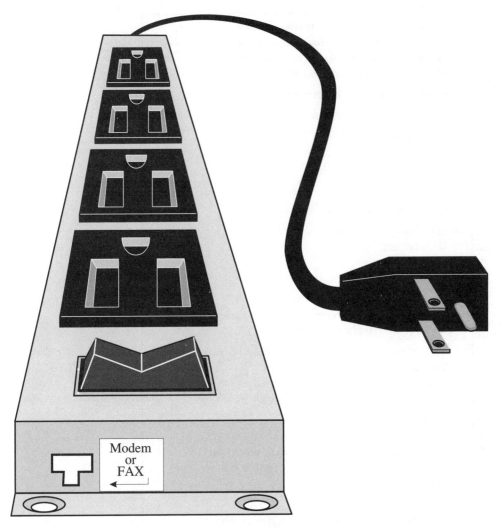

Modem
or
FAX

Figure 4.18 A surge protector on an electric power strip can protect modem or fax equipment from spikes.

8. Put "teeth" into the enforcement program through warnings, suspension of access privileges, or suspension from the job, depending on the violation frequency and severity. Although this sounds extreme, the risks are great plus the consequences of poor security are severe.

9. Ensure that passwords are changed regularly. Make sure they are kept confidential. Many users, if given a choice, will never change their passwords. Also, they will attempt to juggle the same password with a minor change. Require them to use new words and disallow vowels where possible.

10. Determine if the dial modems are available "off the shelf" in case of failures, etc. If a disaster strikes, have a list of modems by speed and type. Create a vendor list showing where new modems can be acquired quickly.

11. Provide lightning arresters and surge protection on all modems. This

❑ Develop a list of users with remote access.
  ❑ Is there a database of users cross-referenced to services?
  ❑ Are multiple log-on's used by same password?
  ❑ Are password denials evident?

❑ Create a system of audit tracking by logging:
  ❑ Date.
  ❑ Time.
  ❑ User ID.

❑ Verify usage.
  ❑ Contact user.
  ❑ Question their usage.
  ❑ Is there denial or confusion existing?

❑ Periodically change modem numbers.
  ❑ Quarterly.
  ❑ Semiannually.
  ❑ Are other exchange numbers available?

❑ Is automatic number identification available?
  ❑ If so, build cross references to ANI and user access.
  ❑ Adds one extra step of security.

❑ Provide an alert mechanism if extended usage exists.
  ❑ Log event by user ID, modem, etc.
  ❑ Notify LAN manager on screen.
  ❑ Provide pager access for off-hours usage.

❑ Enforce security.
  ❑ Write up policies and procedures.
  ❑ Publish and distribute policies and procedures.
  ❑ Have policies and procedures signed by management.

❑ Use corrective measures:
  ❑ First offense: Warning.
  ❑ Second offense: Restrict or suspend privileges.
  ❑ Third offense: Remove privileges or suspend from work.

❑ Change passwords.
  ❑ Frequency: monthly or quarterly.
  ❑ No vowels should be used.
  ❑ Must be a unique word.

❑ Are modems available if disaster strikes?
  ❑ Off the shelf?
  ❑ List type and speed in an inventory.
  ❑ List vendors with modems.

❑ Provide lightning protection.

Figure 4.19 Action checklist of security procedures for dial-in access.

Dial in access list

Name	Dept.	Ext.	Home #	Access to Services	Modem Speed	Priority	Supervisor	Ext.
J. Doe	Finance	5225	800-555-1201	Payroll Finance Server A/P-A/R	1200	4	M. Smith	5001

Restoral Priority: 1= Immediate   3= Medium
2= High   4= Low

Figure 4.20   This form is an inventory of users on the network.

should be done even if the local-exchange carrier has arresters at the entrance to the building. These suppression devices may be simple carbon-block protectors that will not stop high spikes from arcing across the protector and passing through the modem to the LAN. This could be devastating to electronic equipment and network interface cards.

Figure 4.21 is a graphical representation of how the procedures from the checklist in Fig. 4.19 interact. This can be used as a quick reference in a single view.

Dial-out access

As with dial-in access, stringent rules must be enacted for dial-out modems. Before any users are allowed to access modems in an outbound direction, make sure they are fully trained and certified as ready to use the modem communication. Included in this training would be the appropriate authorizations to access specific services. Do not allow random access by *any* casual user. Limit the services that users can access. One way to do this is to build a directory (menu) of numbers authorized to be called, and remove the ability of the user to get dial access. See Fig. 4.22 for a sample of a dial-out access list. Use the same physical protection on all outbound modems as with the inbound modems. A summary of the check points is shown in Fig. 4.23.

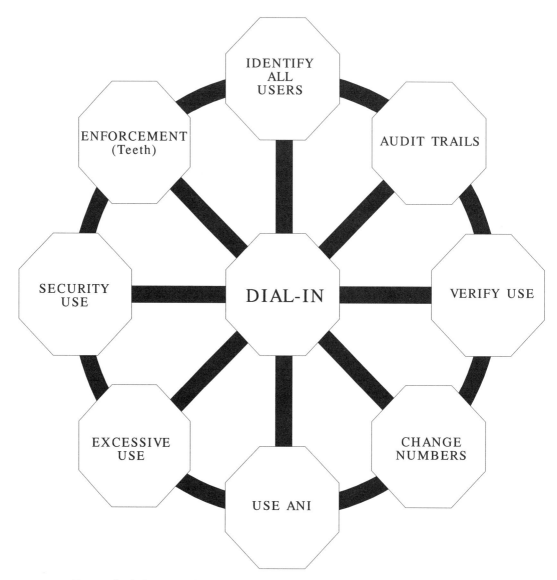

Figure 4.21 Protect the dial-in access by using these procedures.

Modem pools

Occasionally asynchronous modem pools are used on asynchronous communications servers. Periodically check the use of these systems based on the following information, which is summarized in a checklist in Fig. 4.24:

1. Who has access privileges? First and foremost, the privilege is given to a user who warrants it. Access to modem pools should not be treated or considered as a right. This is a service that can be granted or denied based on need. Any user needing access to a modem pool capability must be certified based

Name	Dept.	Ext.	Priority	Services Authorized	Authorized Numbers Dialed	Local Modems Used	Denied Services
J. Doe	Finance	5225	1	Compuserve	1-800-xxx-xxxx	#4 - 1200	
"	"	"	4	Dow Jones	1-212-xxx-xxxx	#6 - 9600	
"	"	"	4	Prodigy	1-800-xxx-xxxx	#4 - 1200	
"	"	"	—	—	1-800-for-game	—	Games BBS

Restoral Priority: 1= Critical   3= Medium
2= Necessary   4= Low

Figure 4.22   Sample of dial-out access list. Periodically check the numbers being called by the users and verify the need. Enforce the security measures as outlined in the dial-in access checklist given in Fig. 4.19.

on the same criteria listed in the preceding sections covering inbound or outbound access. Create a list of access privileges granted to users on the network. In this list, keep a record of the types of services needed, the speeds and protocols used, and the anticipated frequency of use. This database will allow for future updates as needs or technologies change within the organization. Furthermore, this database can be used later to justify the existence of a modem pool, create a list of how often certain services are accessed, create a bill-back arrangement, and verify that only authorized users are availing themselves of the service.

2. What is the duration of usage? How long is an average transmission when the modem pool is used? Is the duration valid? Occasionally a modem communications service fails to break down after a user has logged out of a network or a service bureau (a "hung call"). It would be prudent to understand the actual calling patterns and duration of calls to verify that everything is working properly. A hung call can cost the organization money if this problem exists and proper verification is not used. Furthermore, the hung connection could be an access method for tailgaters to gain access to the LAN. If a modem is hung to the line but no user is on it, the risk of someone coming right in on top of a session could exist. This could allow unauthorized and possibly undetected access to the network. Depending on what status the session was in (active, suspended, etc.) the tailgater could jump right into another user's session. Although this may sound far-fetched, many of the ser-

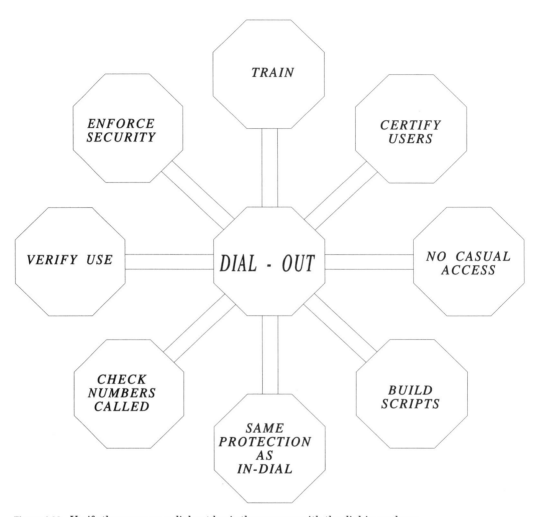

Figure 4.23   Verify the usage on a dial-out basis the same as with the dial-in modems.

vice bureaus have experienced this problem with dial-in users on modem pools. Therefore, it would be a logical extension to look at a LAN access method in the same regard.

3. What are the numbers called? Verify that the users of the LAN modem pools call the numbers that were originally agreed to. If requirements change, that is fine. However, if new numbers appear on telephone bills or packet billing mechanisms it would be appropriate to find out just where the calls are coming from or going to. All too often, security breaks down; this goes undetected due to insufficient verification procedures. Hackers are becoming very proficient in finding new uses of their unauthorized access to LANs such as coming into the LAN from one route (such as a bridge, router, etc.) and crossing the LAN to a modem pool to access some long distance net-

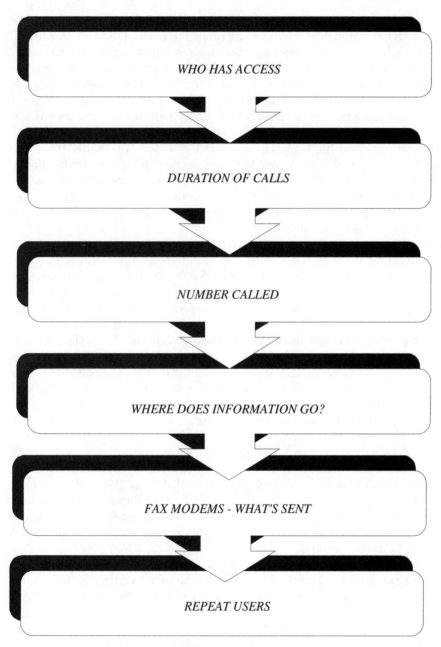

Figure 4.24   Checklist for modem pools.

work, all for free. Using a password on this LAN that belongs to another user makes the penetration less obvious. Assume for a moment that a hacker breaks into a LAN and then accesses the modem pool to dial out onto some other service. The cost of the call is passed to the network owner (organization). However, if that same call is used to break into some other network for fraudulent purposes, another risk arises. The agencies responsible for tracing and trapping hackers may already have equipment on the modems at the far end. Catching the calling number identification, it appears that "your" network users are the hackers. The authorities could conceivably trace the call back to the organization. Armed with search warrants, they could appear at the front door and confiscate everything on the LAN, pending investigation. Could the organization afford this type of publicity? Could it sustain the losses from the unavailability of the computing services on the LAN? How long could the losses be sustained?

If abuse is suspected, act quickly to stop it. Either disconnect the modems, call the appropriate authorities, or change numbers and passwords immediately. The best judge of which action to take is management. Although a purist's solution would be to notify the authorities and catch the perpetrators of this fraudulent use of networks, the timing, cost, and inconvenience factors involved may be more trouble than it is worth. Therefore, management must be apprised of the advantages and disadvantages of this action, so that it can make a valued business decision.

4. Where does the information go after the modem pool is accessed? Can a remote user dial in to the LAN and redirect the data stream to a specific end user on the network? Obviously this opens the door to a remote access through the LAN to specific file servers, files, and directories or any other service on the network. Ensure that multiple levels of passwords are required. This will slow down the random access should a hacker break through front-line defenses on the modem pool. Check what services are readily available to users from modem pools and tighten them up. If a user can gain access, so too can a proficient hacker. Protect these services as they are; a doorway onto your network. No prudent businesspeople would leave the doors to their organization unlocked. So the "doors" to the LAN should be approached in the same way.

Furthermore, if a redirect can be achieved once the LAN is accessed, information that can be copied across the network can just as easily be redirected to the modem port. This applies to both sides of the network, internal and external. Files that can be copied internally could be redirected across the modem line to a remote PC or printer. Information can be "lost" in many ways through these sophisticated access methods being placed on LANs.

5. Fax modems are also a risk. Fax modems and servers offer even newer ways of accessing information on a network. One example of this is obvious: to fax information off the network to a facsimile machine. A user who wanted information from an organization that periodically checked parcel and briefcases at the front door decided to remove the information electronically. A simple phone call to a local business center that provided incoming and outgoing fax

services was made. (Incidentally this can also be sent to hotels, motels, local copy centers, etc.) The user faxed the information to the business center during the day, with a cover sheet advising to "hold for pickup." At the end of the day this person stopped by the business center and retrieved the fax after paying $1.00 per page. What was the value of the information? In this particular case, the fax contained names, addresses, home phone numbers, social security numbers, and payroll information on company employees. This violates personal information and company confidential information. Sold to the correct buyer, this information could reap thousands of dollars for the thief.

This could also cost the organization far more through the loss of employees by corporate raiding, loss of employee confidence if the company cannot protect confidential information, and a general security risk in the industry.

What if the stolen information concerned a new product that was in research and development? What if it was inside financial information that could have a significant impact on the company's competitive position or stock value? Secure the fax as you would secure any transmission service on the network.

6. Periodically verify repeat users and the location being called. This is administratively troublesome but is worth the time and effort. Let users know this will be done. They will probably rebel at first so educate them on the risks associated with the communications access and the potential losses exposed here. Attitudes tend to shift when users understand the security risks involved rather than an image that "Big Brother is watching" everything they do.

Dial-back modems

Many LAN administrators install dial-back modems as a security measure, thinking that these systems are a cure-all for their risks. True the dial-back capability does offer additional protection, but the proper devices must be selected. Too often breaches have occurred due to a lack of understanding of how these devices work. Early versions of dial-back modems (and many still exist on LANs and other data processing functions) used a simple procedure given in the list in Fig. 4.25. Use Fig. 4.26 as a simple reminder of how the flow works. Display it in front of the modem.

Figure 4.25 reflects the way the dial-back modem should work. A note of caution here: the password for users should be as long as possible (preferably seven digits or more). However, users always want it to be as short as possible. Compromises on password length to satisfy a user only increase the risk of security problems.

A problem exists with the dial-back modem, particularly the earlier versions mentioned above. Many of the older versions are fairly easy to break through. These devices do not understand dial tone or answer-back supervision. Therefore hackers have been able to penetrate a dial-back system even though the LAN administrator believes everything is secure. How this is accomplished is summarized in Fig. 4.27.

1. A database is established of the authorized users of the modem, consisting of the user name, user ID, home number (fixed), and password.

2. The authorized user dials into the network modem, which answers and goes to silent tone. The modem waits for the user ID and password.

3. Upon entering the required information, the user then waits.

4. The modem looks up the user ID and password for authorization. If a valid user ID and password have been entered, the modem sends back a confirmation tone. If an invalid user ID or password are entered, the modem will either send back an error tone or cut the caller off.

5. The modem waits a period of time (3 to 5 seconds), then seizes the line and dials-out the fixed number (home) as established in the database.

6. The user receives an incoming call and allows their automatic answer modem to go through a hand-shake with the dial-back modem.

7. The user then hits two carriage returns, signaling the dial-back modem that everything is alright.

8. The dial-back modem then allows the user access to the LAN or computer services on the network. Everything works just fine.

Figure 4.25 Procedure for how dial-back modems work.

At this point the casual user will think that there is no problem. Since the modem is going to call back the authorized user, the hacker will not receive the call back and therefore will be denied access. The system works.

This is true except that variations in the way this works could be used. The authorized caller, after receiving the confirmation tone, is supposed to hang up so that the modem can dial back. The list in Fig. 4.28 is a continuation of the process after the hacker breaks a password (Fig. 4.27).

There are ways to prevent this breach from happening. First, check to see if this condition exists with the modems in place. If so, replace them immediately! Other actions could be as follows:

1. Use a kill program that will shut down the modem after three failed attempts to get on the modem. Since it will take the hacker several attempts, the kill program will help to prevent access.

2. As failed attempts are made, use a dial-out sequence to a security person, pager, or systems administrator's terminal. This will alert the administrator that someone is attempting to break in.

3. Use a modem that will force a disconnect of the caller after a valid password is entered. After the disconnect the modem will seize the idle line and call the user whose password was entered, regardless of whether the actual user or a hacker entered the password.

4. Educate users that if they receive calls off-hours from a modem to report it immediately. They should be aware that this could signify that someone has broken through the system using their password.

5. Install a modem that uses two separate lines. If a valid password is entered, the modem will seize the second line and dial-out the appropriate user whose password was entered. A hacker can sit on the first line all night long but the modem will not let him or her in.

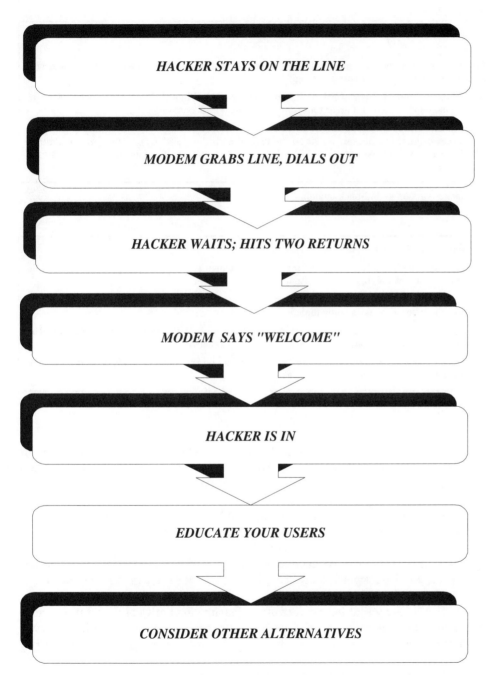

Figure 4.26   Breaches of the system can be caused by dial-back modems.

1. Using an older modem and a single line for incoming and outgoing calls, the user sets up a call-back arrangement.

2. The hacker dials into the modem repeatedly and attempts to use a valid password. The modem will look for a fixed-length password (assume six digits).

3. As the hacker begins entering a password, a digit at a time issued. When the hacker gets six digits into the system, the modem will send an error tone, or cut the hacker off.

4. Now that the hacker has learned the length of the field, six digits, the use of PC programs to generate random sequences begins. The hacker can sit on the line all night long and keep trying (or use an automatic dialing program to do this) until a successful password is entered. The modem then gives a call-back confirmation tone.

Figure 4.27   The hacker version of dial-back modems.

5. Instead of hanging up, the hacker stays on the line and waits.

6. The modem goes through its cycle of waiting for 5 seconds. It then grabs what is supposed to be an idle line and begins pulsing out the digits of the intended authorized user.

7. Since the hacker never hung up, the dialed-out digits come across the line through the modem speaker. The hacker actually hears the digits being dialed.

8. The hacker waits a few seconds, then sends an answer-back tone to the call-back modem. After the handshake occurs between the modems, the hacker hits two carriage returns.

9. The modem allows access to the network. Security is breached and the audit trail shows that an authorized user was on the system, which obviously was not true.

Figure 4.28   The breach occurs.

6. Consider testing all dial-back modems. Whether a modem with one or two lines is used, test it. Dial into the system, enter a valid password. Then wait and see what happens. If the modem pulses out the digits on the line in use, change it immediately or turn it off.

7. Look at other alternatives such as ANI, which can be used as an additional security check. If the modem can pass the ANI digits through to a central processing unit (CPU) to verify that the call is from an authorized telephone number, this can be used as a transparent security check to the caller. Even if a valid password is entered, if the ANI does not match the database, access can be denied.

8. If ANI is available, use a capture feature for both valid and unsuccessful attempts. This can be used as an audit trail to trap the calling party's number. Report violations immediately to your appropriate carrier's security department.

9. Look at other alternatives to prevent unauthorized access. Although more expensive, changing key authentication services as provided by companies such as Lee Mah Systems or Security Dynamics will add security to the LAN.

10. Change passwords and phone numbers on a regular basis if failed attempts are logged. This means that someone is attempting to bypass the security system and access the LAN. Again, the inconvenience and costs

associated with this choice are dwarfed compared to the losses possible from a breach in security.

Use Fig. 4.29 as a check sheet to educate users on what their expected reaction should be with dial-back modems. Furthermore, use the alternatives summarized on this sheet.

Leased-line modems

Protecting leased lines is a bit more difficult, since the external part of the network is beyond your direct control. Usually this falls in the domain of the

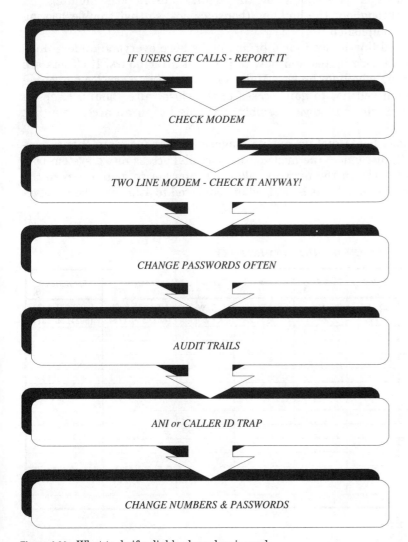

IF USERS GET CALLS - REPORT IT

CHECK MODEM

TWO LINE MODEM - CHECK IT ANYWAY!

CHANGE PASSWORDS OFTEN

AUDIT TRAILS

ANI or CALLER ID TRAP

CHANGE NUMBERS & PASSWORDS

Figure 4.29 What to do if a dial-back modem is used.

local-exchange or interexchange carrier. However, at the entrance facility (cable vault) check to ensure that lightning arrestors and spike eliminators are installed on the cable pair. A transient spike of electricity can severely impair your LAN by blowing out servers, terminals, printers, etc. At the demarcation block make sure the local-exchange carrier (LEC) has tagged the circuit as mission critical or put red markers on the block.

Other things you can do include the following.

1. Make sure you have all the information on the line for repair purposes. This includes the items listed in Fig. 4.30: Refer to Fig. 4.31 for an explanation of the information contained in Fig. 4.30.

Once this inventory is conducted, create a graphical schematic of the circuits as shown in Fig. 4.32. This shows how the circuit comes into the building and where it runs throughout the building (house cables) so that troubleshooting can be easily accomplished.

2. If the leased line is used for a bridge make sure everyone understands where the circuit terminates and how to get at it after hours. If closets are locked, determine who has the key. Make sure appropriate service personnel are on an authorized list to gain access to the closet after hours. Keep an escalation list of who has keys. Possibly post this list in an accessible (but secure) area.

3. If the bridge has a dial-up remote diagnostics port for maintenance personnel to test it, disconnect the modem. Random dial access into a system port may lead to a breach on the network, allowing a hacker to gain access to the LAN and to any other LAN it is bridged or connected to. This may be inconve-

LEASED-LINE INVENTORY							
Circuit ID ①	Cable Pair ②	PIN # ③	Jack # ④	Block # ⑤	Carrier ⑥	Repair # ⑦	Escalation # ⑧
67TL157286CD	B6P101-102	22 - 25	7 - 126	4 - 108A	U.S. West	(602) 555-1000	(602) 555-1333
624WD5816	"	"	"	"	AT & T	(602) 550-1234	(602) 550-1811

Figure 4.30   Leased-line inventory.

nient, since any access to the bridge maintenance port will require a staff member's intervention to plug in the modem. However, this is well worth the effort.

4. If a network employee leaves the company, change telephone numbers, passwords, privileges, etc. immediately. Do not assume that nothing will happen. The circumstances surrounding this departure may be less than amicable, leaving a disgruntled former employee and a security risk at the door to the network.

5. In the event of the loss of building access, have a plan in place to reestablish the leased line at a hot site, shared site, or whatever. Consider the option to have a reserve link ready at the new site in advance. However, this may be costly. A cost–benefit justification should be used before arbitrary expenses are incurred. The justification must weigh the monthly costs against the value of data, access, and productivity of the connection.

6. Consider an automatic dial-backup capability at the bridge (or other point) in the event that the leased line fails. This will be stored in memory to a single predefined modem number. Keep this number confidential!

7. Be cognizant of any construction in and around your building. Construction will inevitably lead to backhoe fade or a cable cut that will interrupt the circuit, thereby impairing connectivity from LAN to WAN or LAN to LAN.

8. Use closet diversity inside the building when distributing cables on the house cable pairs. Dual risers and closet diversity, as shown in Fig. 4.33 prevent single points of failure from internal cable cuts, floods, fires, etc., that are isolated on one side of the building.

---

(1) Circuit identification is provided by the carrier. If a cross-connect is used by the local-exchange carrier (LEC) to an interexchange carrier (IEC), the IEC's circuit ID should be referenced on the inventory. This can be listed directly under the original circuit number, showing the same cable, pin, jack, and block numbers. The responsible carrier (i.e., controller of the circuit) should be listed first.

(2) Cable pair of the circuit running into the building from the LEC: use cable number, bundle number, and pair number sequence.

(3) The pins that the circuit is terminated on—in this case a four-wire circuit is used, so 4-pins are referenced.

(4) The jack number refers to the RJ-21X; in a multijack environment the jack number is 7 cross-connected to a house bundle pair number 126.

(5) The block number (4-108A) refers to the interface inside the building, such as an RJ-45 jack located on the fourth floor, in office 108, location A.

(6) Carrier refers to the appropriate company to call depending on who bills for the circuit.

(7) List repair service phone numbers to call if a circuit goes down; this may include a daytime number and an off-hours number.

(8) Escalation numbers are used if the repair phone line is busy or if the circuit is down for longer than a predetermined and agreed to period of time (i.e., 15 min). The escalation is used to get critical circuits back quickly. Usually this is an operations supervisor. The escalation column may also include an off-hours (home) number for the supervisor.

Figure 4.31   Key to information in Fig. 4.30.

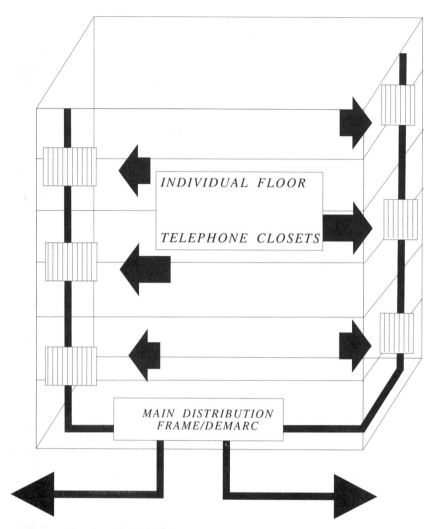

INDIVIDUAL   FLOOR

TELEPHONE   CLOSETS

MAIN   DISTRIBUTION
FRAME/DEMARC

Figure 4.32   Mapping out the facilities.

Communications security

Above and beyond the protection items listed above you should also consider the following areas, which are summarized in Fig. 4.34 as a checklist.

1. Access methods: What ways or combinations of ways can be used to gain access to the LAN? Check them all and secure them quickly.

2. When random access attempts are made on a single modem or a modem pool, ensure that the ability exists to get a count of attempts made, whether successful or failed. This provides a good audit trail that should be checked and verified routinely.

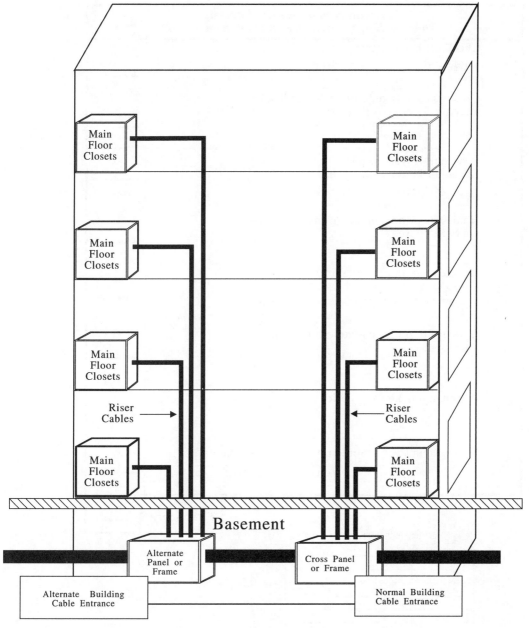

Figure 4.33 Closet diversity reduces the risk of single points of failure inside the building.

3. Password protection should be several layers deep. Do not allow users to use names or birth dates for their password. Use an authentication key where possible. Preferably a changing key method that uses a new password after every use should be selected. Consider fixing the password with a combina-

Communication Security Checklist *What levels of security risk exist on the network?*	*Score*
① Check methods of access used: (For each check score one point) ☐ Dial-In (1) ☐ Dial-Out (1) ☐ Call-Back (1) ☐ Leased (1) ☐ Bridged (1) ☐ Routed (1)	_____
② Do tools exist to measure communications attempts ? ☐ Yes (1) ☐ No (2) If attempts can be counted, does system count failed ☐ or successful ☐ ??	_____
③ How many layers of password exist ? ☐ 1 ☐ 2 ☐ 3 (3) (2) (1) Do provisions exist to restrict certain types? ☐ Yes (1) ☐ No (2)	_____
④ If mission critical applications are on LAN, encryption used? ☐ Yes (1) ☐ No (2)	_____
⑤ Is watchdog timer available ? ☐ Yes (1) ☐ No (2) If yes, how long is it set for? _____ (Score)	_____
⑥ Is "Kill Program" available for failed attempts ? ☐ Yes (1) ☐ No (2)	_____
⑦ Are changing key authentication codes used ?	_____
⑧ Will system alert administrator of failed attempts ? ☐ Yes (1) ☐ No (2)	_____
	Total Score _____

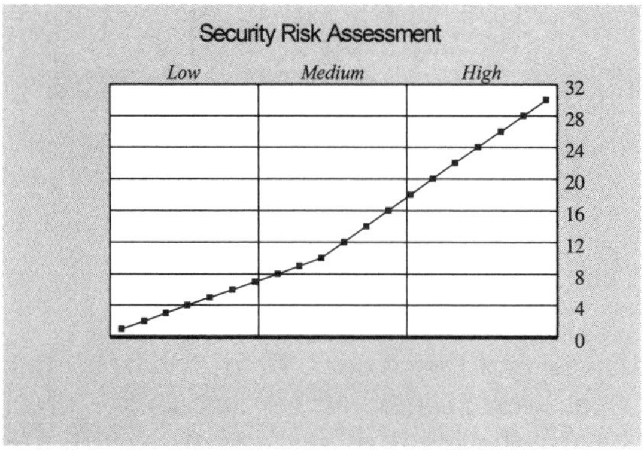

Security Risk Assessment

Low   Medium   High

32
28
24
20
16
12
8
4
0

Figure 4.34 Communications security checklist.

tion of alphabetic and numeric digits. Also consider using *, #, !, etc., digits as part of the password to exponentially increase the combinations available.

4. Encrypt passwords and the data (mission critical, payroll, finance, strategic plans) whenever possible. A breach of a LAN will help to protect your data if encryption is used. Make sure that LAN analyzers do not exist on user workstations (e.g., EtherPeek, Lanalyzer, Spider, Sniffer, etc.) since these could override security systems.

5. Use watchdog timers so that a user connected to a communications device (modem, bridge, etc.) after three, five, or ten minutes of inactivity will be logged off the system. An example of this is provided by Citadel Software Systems in a product called "Net-Off." This TSR application will log the user off after a predetermined period of inactivity.

6. In the event of failed attempts, use a kill program that deactivates a modem (or other communication device) and leaves it off until manual intervention by a LAN administrator. Follow up on users who failed, and find out why. Make sure it was the actual user, not a hacker. When the kill program is activated, consider using an audible alarm or a dial-out sequence to LAN personnel, requiring immediate action.

7. *Never assume that communications are safe! All communications systems and subsystems are penetrable, it is only a matter of time before security is breached.*

Chapter

5

Physical Recovery

Physical recovery in this book will look at the potential points of failure on the network. These are varied where combinations and permutations exist. However, a recovery process can only be addressed within certain guidelines during the planning process. If an attempt to address every single combination of failure was made, the list would never end. Instead, a realistic approach should be used, in that general categories should be considered. Upon categorizing the major events, a flexible plan can be used to address the variations. Whether a cable system is damaged by fire, flood, human error, or other destructive acts really does not matter. The real problem is that the cable is unusable. Therefore, this general listing by category can be considered. Thereafter, the scenarios can be played out in a plan or in a LAN manager's operational manual. The choice is one that belongs with each individual. This chapter addresses damages that can occur to the following major physical equipment categories. See Table 5.1 for a summary of these risks, outlined in areas that may require physical recovery procedures.

1. Cable system
2. Media access units (MAUs)
3. Servers
4. Workstations
5. Terminals
6. PC or host

Each of these components is an integral part of the LAN and must be addressed in a logical order. The first three on this list can cause significant downtime to the entire network, whereas the latter three will affect individuals. Regardless of which category suffers failures, the net effect is that the users are impaired from doing their jobs. The possibilities of damage cover so

TABLE 5.1   Areas to Check for Physical Risks

Item	Risks
Cable systems	Cuts, internal or external Fire damage Water damage Rodent damage Electromagnetic or radio frequency interference
Media (multistation) access units (MAU)	Droppage or other physical damage Electrical risks from spikes Overheating Theft
Servers	Fire and smoke damage Water damage Physical damage from being knocked over Electrical spikes Loss of power Tampering Loss of building or loss of access
Workstations	Fire or smoke damage Physical damage from droppage Beverage spillage by users Water damage Power up failure Unavailable access to network
PCs and host systems	Fire and smoke damage Water damage Physical damage from droppage Power spikes, sags, dips, etc. Theft Unavailable access to network or building loss Incompatible cards causing conflicts

many combinations that it would be best to play out different scenarios. These will include the following.

Cable systems

1. Cable cuts—both internal and external.

2. Cables burned from fire—either in the organization's area or anywhere else in a multitenant building.

3. Cables flooded from water damage—damage can come from ground level, sprinklers, or other areas above the area.

4. Protecting against rodent damage—general category: mice, rats, gophers, etc. Other species can also fall into this category depending on geography (i.e., fire ants).

5. EMI and RFI interference—the cable system can become a large antenna, drawing in interference from electrical sources or radio frequency sources.

Either way, the impact is the same: disruptions and corruption of data and files on the LAN.

Media access units

1. Units dropped—this can be minimized with proper training, but installers still must be made aware of the problems that will arise if a suspect MAU is installed.

2. Units hit with electrical spikes—a large amount of problems occur from poor electrical protection.

3. Units overheating—although not as great a problem in the past, newer indications point to improper conditioning of closets, causing loss of units due to temperature extremes.

4. Robbery of units—theft of these units is on the rise; they must be protected as much as any other corporate resource. Thieves do not realize what they are stealing, but since the MAUs are centralized and have electronics, the temptation is often too great.

Servers

1. Destruction by fire or smoke—servers are PCs or minicomputers and are prone to the same damaging effects of fire, heat, and smoke. Proper recovery procedures are required after damage.

2. Physical damage—getting bumped or knocked over. This is becoming more of a threat than before. As more servers are added, the space becomes cluttered, increasing the possibility of damage.

3. Inaccessibility to network—cannot log on. As damage occurs, the server may become inaccessible. When users cannot log on, the damage and disruption are magnified exponentially.

4. Building unavailable—what if you cannot access the building to properly maintain and back up the server? Provisions must be made in advance to overcome this problem.

Workstations

1. Destruction by fire or smoke and water—all of the risks associated with the above components must also be considered with workstations.

2. Physical damage—getting knocked over; spillage. Users can be careless and cause many areas of disruption. Some of these include knocking the unit over; droppage due to arbitrary movement. Others involve spillage of drinks on the units.

3. Failure to power up—power spikes or other potential risks exist at the workstation. Simple problems of units being unplugged can also occur.

4. Inaccessibility to network—if a workstation cannot access the network, it is virtually useless. This can combine with other problems listed above.

PCs and host systems

1. Fire, smoke, or water damage—these devices are sensitive to destructive forces; replacements or cleanup actions must be thoroughly planned to prevent future occurrences.

2. Physical damage—getting knocked over or bumped. As with workstations, PCs are exposed to user damage. Hosts attached to the LAN are better protected by the IS groups. However, a host problem could be greater in consequential risks.

3. Power problems—power is becoming one of the single largest risks involved with electronic equipment. Proper recovery involves proper power protection. Furthermore, PCs use standard power supplies. As users add more internal cards to accommodate desires, they strain the power capacity.

4. Unavailability—whether from theft of the unit or from the loss of access to the network, proper recovery of company assets must be considered.

Cable Systems

Many of the problems users will experience with network downtime will be the direct result of the cable system. This is an expensive piece of the network (the average cable pull to a user is $250 to $400), yet it usually gets overlooked. It is also the single most common opportunity people use to cut costs. A recent study indicates that cable failure is a major cause of network downtime, costing companies $3.5 million in productivity and $600 thousand in network downtime. For the few dollars saved by cutting corners, the resultant downtime, disruptions, and errors are introduced. Therefore, recognize that the cable is a long-term investment and that depreciation on cables will spread the costs over a greater period. There is no real justification for cutting costs. Remember that the use of the cables for local area networks is designed around bursty, high-speed data. Therefore when planning a cable system you should follow the guidelines listed in Fig. 5.1.

In many cases, the cost of increasing the level of cables to accommodate future growth is minimal. This is a decision that must be made in advance of the installation. One recent small network of 20 stations increased the level from 3 to 5 and had a net increase of $150 to do this (or $7 per station pull). This is hardly a major penalty when compared to the potential cost of rewiring each run at a later date. The difference is summarized in Table 5.2. Management must be made aware of the business impact and long-term growth projections for the network and the cost implications of doing it wrong, if this decision is to be supported.

Think of the impact if the 20-station network described in Table 5.2 was increased to 2000 nodes. The pricing would be proportionately different.

1. Do not attempt to use existing telephone wiring in a building. This cable will not support the high-speed data needs of the LAN.

2. If using unshielded twisted pair wiring, stay with (at a minimum) the underwriter laboratory (UL) and Anixter adopted levels of wiring applicable for your LAN.

3. For a token ring network operating at 4 Mbits/s use level-3 cable, 24 gauge, solid conductor for up to approximately 100 m (328 ft).

4. For 10 base T networks operating at 10 Mbits/s use at least level-3 cable, 24 gauge, solid conductor for up to approximately 100 m (328 ft).

5. For a token ring network operating at 16 Mbits/s use level-4 cable, 24 gauge, solid conductor for up to 100 m (328 ft).

6. For high-speed networks operating at up to 100 Mbits/s [such as fiber distributed data interface (FDDI) on twisted pair, or now copper distributed data interface (CDDI)] use level-5 cable, 24 or 22 gauge, solid conductor for up to 100 m (328 ft).

7. If installing new networks at 4, 10, or 16 Mbits/s consider using level-5 cable throughout. This allows for future growth (speeds) without the need to rewire the building at a higher cost later on.

8. Any rewiring due to moves, adds, changes, or repairs should be consistent with the above migration strategy. Do not allow cable contractors to substitute lesser-grade cables.

9. Patch panels at closets and crossover cables from patch panels to MAUs, hubs, or concentrators must use the same quality cables. Do not use flat ribbon cables. Use level-3, -4, or -5 (based on the cable plant) twisted pair cables. See Fig. 5.2 for a graphic representation of this component.

10. Drop cables from the wall outlet to the workstation (PC, server) must be the same quality cable. Use level-3, -4, or -5 drop twisted pair cables (depending on the cable plant). See Fig. 5.3 for a graphic representation of this drop cable.

11. If the number of twists per foot for cables must be specified, use 40 twists per foot to minimize interference and cross talk. See Fig. 5.4 for a graphic representation of the number of twists per foot.

12. Make sure all connectors (i.e., RJ-45 jacks) are made for the appropriate cable thickness and levels being used (see Fig. 5.5). An undersized connector will cause potential problems on the network due to poor crimps, loose connections, or broken or split connectors.

13. Verify everything by physically inspecting these connections.

14. Keep prepared proper connectors and crossover and patch cables on hand for spares and testing.

Figure 5.1   Decision making for cable runs on LANs.

TABLE 5.2   Cost Comparison of Wiring Level Increase Today vs. Rewiring Later

	Present wire increase		Future rewire alternative
Cost per drop	$250.00	Cost per drop	$250.00
Number of drops	× 20	Number of drops	× 20
Cost of installation	$5,000.00	Cost of installation	$5,000.00
Cost to add higher-quality		Cost to add higher-quality	
and -level cables	+ $150.00	and -level cables	+ 0.00
Subtotal	$5,150.00	Subtotal	$5,000.00
Rewire 20 drops in future	+ 0.00	Rewire 20 drops in future	+$5,150.00
Total	$5150.00	Total	$10,150.00
Delta cost	$150.00	Delta cost	$5,150.00

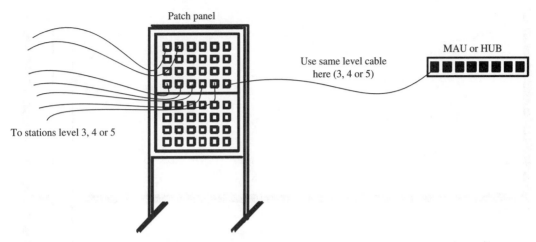

Patch panel

MAU or HUB

Use same level cable
here (3, 4 or 5)

To stations level 3, 4 or 5

Figure 5.2   Use the same level cable for patch panel and crossover cables.

First and foremost understand how the cables are run throughout the building. A detailed cable plan should be generated so that the cable number, pins, and jacks associated with each other are all known. The existing cable plant inside a building is often used to run drops from a closet to the workstation or on the riser side to get from floor to floor. Although this may work fine for some networks, the existing plant is usually undocumented. Compounding this problem, the older cables are frequently used for telephone wiring, which means that excessive taps, wear, bridging, and abuse probably all exist. Through a detailed inventory of the cable plant, many potential pitfalls that are just lurking in the periphery of good LAN operations may soon be discovered.

- *Use a database* to document every cross-connect within the environment. This was discussed in Chap. 4, but the significance becomes even greater when trying to recover a workstation, server, or a segment on a LAN. The database makes it much easier to trace problems from the documentation available. The database could be put into a relational form (using some of the more popular software packages on the market such as Oracle, Paradox, etc.) or in a spreadsheet (i.e., Lotus, Excel, or Quattro) for ease of use, in changes and printing the sheets. A paper-based system can also be used but may erode quickly in a dynamic environment. Figure 5.6 is a representation of this database.

- *Label every cable* (large bundle) by number to a cross reference. Every cable should be numbered on both ends at a minimum. Preferably the labels should be placed every 10 to 20 ft so that tracing can be accommodated. See Fig. 5.7 for this labeling scheme.

- Show the routes (either on a schematic or in text to describe how the cable runs from location to location). The preferred method is a graphic, such as

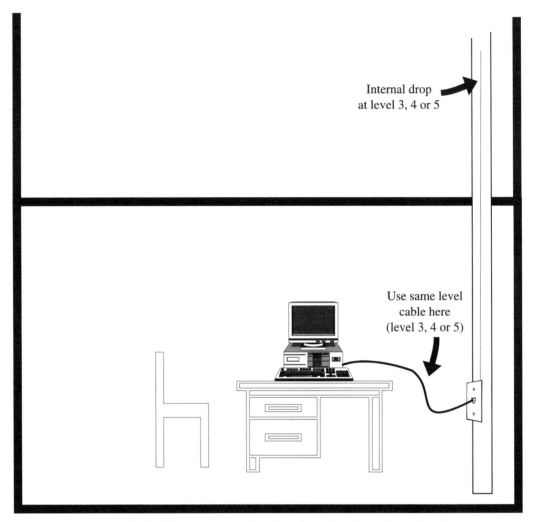

Internal drop
at level 3, 4 or 5

Use same level
cable here
(level 3, 4 or 5)

Figure 5.3   Use the same level cable at the station drop from the wall outlet to the workstation.

a computer-aided design (CAD) file. However, this can be done on various cross-referenced forms. These should be as detailed as possible so anyone can pick up the reference document and recover cables. Keep in mind that after a major disruption or disaster, chaos will reign. As much documentation as possible will assist in having others (i.e., installers, vendors) offer their assistance in repairing or replacing a cable plant. This is of particular importance after cable cuts, fires, and floods since major portions of cables may have to be replaced.

An old technique of using the damaged cables as "pull strings" will not work here. Since cable trays and conduits get packed very quickly, trying

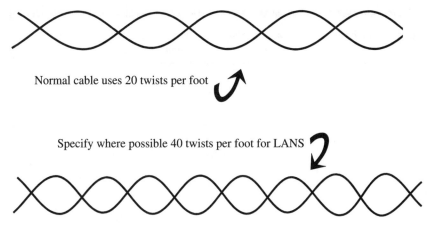

Normal cable uses 20 twists per foot

Specify where possible 40 twists per foot for LANS

Figure 5.4 Specify where necessary 40 twists per foot.

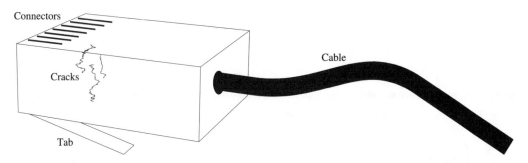

Figure 5.5 Use the correct jacks for the size cable used. When crimping the cable onto a jack, the wrong gauge cable can cause cracked connectors, loose connections, or improper crimping. Any of these will cause problems on the network.

to use a damaged cable that is buried in a heavy bundle can cause further disruptions, as follows:

1. Hard tugging can break the pull cable.

2. Resistance by other cables and the subsequent tugging could strip the insulation off many surrounding cables, causing extensive additional damage.

3. Many cables have been greased to pull them in originally. Over the years the grease may dry out, causing bundles to become caked together like concrete. Using one of these cables as a pull string may damage the surrounding cables.

Floor	Cable #	Drop #	Patch Cable	Hub Crossover	Hub	User Location	Jack	Device
2	156	A101	3-23	1-7	4	2-101	17A	PC

Figure 5.6 Keep cable records up to date and as detailed as possible.

- *Isolate the cables* from transformers, motors, and other noise-producing systems.The best way to do this is to use conduits or flex tubes. However, as a cost measure, most cables are run open in ladders or trays. These open runs may be dragged across fluorescent lights. When cables are dragged across or laid on top of lighting fixtures the cable draws in the hum (noise) onto the copper. This is electromagnetic interference (EMI), which was previously discussed in Chap. 3. To further complicate this process, many cables are run in very diverse methods. Often, elevator shafts have been used for risers, etc. Shafts are both a high-noise area from the electric motors on the elevators and a dirty environment in which to run the cable.

  Another potential problem is the use of electrical closets positioned in the office space as wiring closets. These electrical closets have high-voltage transformers and electrical equipment that can generate significant EMI on a cable system. Every wiring specification has certain distance limitations (at a minimum 6 to 12 in) from these high-noise-producing pieces of equipment. If the proper distance cannot be achieved, consider using J hooks hung from beams in the ceiling space. The J hook allows for cables to be hung from the joists at the necessary distances. See Fig. 5.8 for a representation of the use of J hooks.

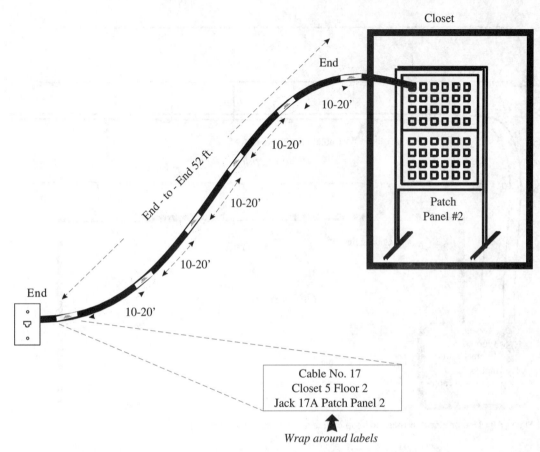

Closet

End

10-20'

10-20'

End - to - End 52 ft.

10-20'

10-20'

10-20'

End

10-20'

Patch
Panel #2

Cable No. 17
Closet 5 Floor 2
Jack 17A Patch Panel 2

Wrap around labels

Figure 5.7 At a minimum, label every cable on both ends. Labeling at 10 to 20 ft intervals allows easier tracing and troubleshooting.

- *Shielding,* if used, should be checked to ensure that it is grounded on one end only to prevent the existence of ground loops. This can be a major problem. The shielding is used to prevent EMI and radio frequency interference (RFI) problems. However, when installed, there may be two different installers pulling or terminating the cable at different ends. When terminating the cable the installer may install the two- to eight-conductor cable onto the appropriate block or patch panel. Finding a stranded extra wire, the installer will carry this to ground. This is correct, as it is the purpose of the shield to carry voltages and noise off to ground.

    However, at the opposite end a different installer may go through the exact same procedure. Herein is where the problem occurs. Now that both ends of the cable have been grounded, two different ground potentials may exist. This creates a closed circuit that will carry electrical current right on the cable system. The cable has become an antenna, drawing in noise and RFI. Causing a ground loop draws noise onto the LAN, which will cor-

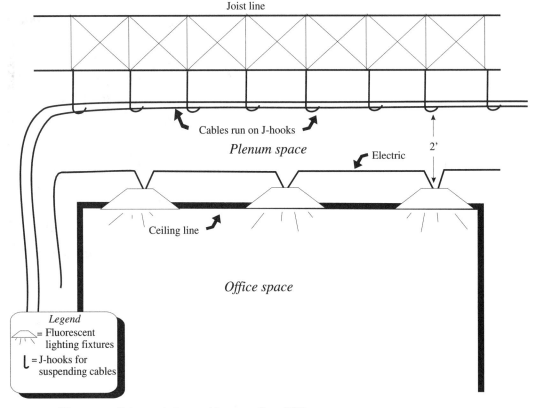

Joist line

Cables run on J-hooks

Plenum space

Electric

2'

Ceiling line

Office space

Legend
= Fluorescent lighting fixtures
= J-hooks for suspending cables

Figure 5.8 Use proper distances to keep cables away from EMI sources.

rupt the data traversing the cable system. See Fig. 5.9 for how this might happen using the ground loop on a shielded cable system. A further complication to this ground loop risk is the potential for transient voltages from lighting, spikes, etc., to be carried across the ground and literally expose the equipment and users to danger from this voltage. A good spike can "fry" a PC, server, and many of the components along the line. Cut one end of a ground loop to solve this problem.

- *Color code* conventions must be maintained consistently throughout the area. Only high-grade cable (24 gauge or better) should be used. The color code and pin-outs for jacks have been defined and accepted by the industry. This allows for the proper connections and signals to be placed on the appropriate wires. However, from time to time problems arise. If conventions are not adhered to, various possibilities exist. The most critical will be delays in recovering the network, if an installer has used a different color code based on multiconductor cables. The repair process can be delayed significantly. Troubleshooting on the wrong wire will produce incorrect results, thereby causing confusion.

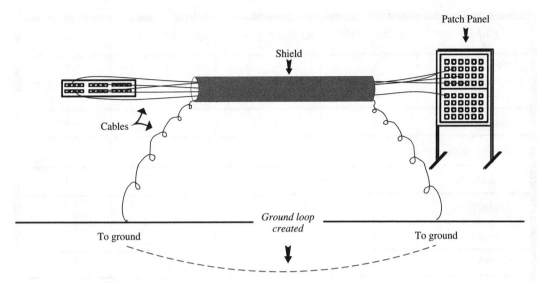

Figure 5.9   Make sure only one end of a shielded cable is grounded.

The industry convention for wiring is to use the wires by paired colors (one wire being *transmit,* the other *receive,* etc.). A typical example of this is shown in Fig. 5.10 for a four-pair cable (eight conductor) and a Decconnect[1] six-conductor wiring scheme.

As evidenced in Fig. 5.10, the wiring of the pin-outs differs, depending on what conventions are being used. If installers use conventions they are comfortable with, the risk is always present that in a disaster recovery process, the wrong wiring pin-outs could be used by the new installer. Furthermore, cases have been recorded where an installer uses any convention that deviates from the wiring conventions. This is either from unfamiliarity with standards or a form of job security. Regardless of the reason, the system will take longer to recover if an event causing failure occurs and wiring must be rechecked along the entire run of a connection.

Bear in mind that recovery procedures of a network must be clear and precise enough so that anyone can pick up the pieces after a disaster. By maintaining conventions this will be more easily accomplished.

- *Backbone cables* such as fiber or coaxial should be labeled at each closet to isolate specific runs. The route and any repeaters used should be checked for proper connections and for adherence to distance limitations. The distance limitations are stated for the best transmission possible. Deviating from these standards can cause problems. When recovering a network, the

---

[1]Decconnect™ is a registered trademark of Digital Equipment Corporation.

Color	RJ45 USOC	RJ45 10 Base T	RJ45 TIA-568	MMJ DEC
White/Blue	5	5	5	3
Blue/White	4	4	4	2
White/Orange	3	3	3	4
Orange/White	6	6	6	5
White/Green	2	1	1	1
Green/White	7	2	2	6
White/Brown	1	7	7	
Brown/White	8	8	8	

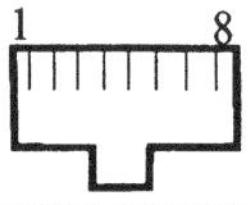

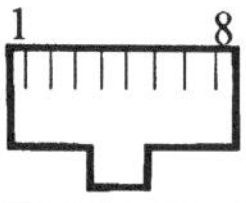

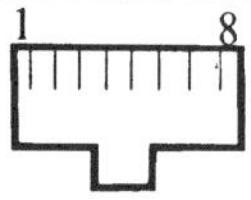

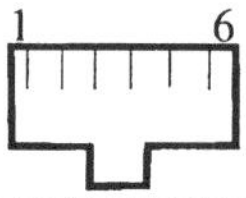

Figure 5.10 Four-pair color code wiring.

cleanest transmission possible should be achieved. If, however, thresholds and signal quality are already impaired, splices necessary to recover may cause the network to degrade. Obviously, the best and strongest signal transmission should be the goal, but short-term fixes may be required. Therefore a schematic should be used to show the following:

1. Distances of cable runs in the backbone

2. Repeaters and locations

3. Ground loops, if any, on the backbone

4. Signal levels along the run

5. Splices if any (or barrel connectors)

■ At *patch panels* or *cross-connect points* all connections should be checked to ensure that the proper seating of the wires is achieved. Patch cables can be enigmatic. First, physically check to see if any visible damage exists on the connector. Second, use a spare cable that has been tested and proven that it works. Third, if all else fails replace everything. These cross-connects at patch panels are constantly being plugged and unplugged. Damage can occur, so they must be checked regularly. If an intermittent network problem occurs, the likelihood of a patch cable being

1.	Check that the patch cords do not have excessive strain. If a cable is too short, replace it with a longer one.
2.	If stress is placed on the cable from the weight of others draped on it, the risk of the conductors pulling away from the connector is high.
3.	Relieve strain from the weight of other cables by tying these down to a solid surface or by adding stress relief coils.
4.	When plugging in or removing cables, make sure that the plastic connector is grasped firmly. Do not yank these by the cable.
5.	Never allow anyone to drape anything on the patch cables.
6.	When any patch cable is in doubt, replace it.

Figure 5.11   Checklist for cross-connects.

damaged or loose is high. Use the check points in Fig. 5.11 to verify cross-connect on cable systems.

- *Check connectors* on all copper, coaxial and fiber cables for proper connection. Just as patch panels are important, the connectors on all cables are equally important. Use prudence on all connections. If they are in doubt, replace them.

- Once this *inventory* is complete, it is imperative that it is kept current with all move, add, and change activities. Preferably this will be done in an electronic format. Use a database that is easy to keep current. The best system is one that can be quickly and conveniently updated. Verify this at least monthly by conducting spot checks. After any new installations or moves, set a time frame during which the data will be entered. Stick to this time frame or the system will erode and become obsolete quickly.

Cable cuts

Should a cable cut occur, anyone should be able to resolve the problem quickly. Armed with detailed schematics and the inventory as outlined above, begin to isolate the problem, then fix.

1. If the cut occurs on the riser, the splicing effort will be the most time-consuming task. However, when splices are introduced, a degree of loss may also be introduced onto the cable. Make sure that the splice does not exceed specifications for loss on the cable plant. See Fig. 5.12 for a graphic representation of a backbone cable cut.

2. If the cut is on an individual station drop, a whole new drop can usually be run within a reasonable period of time. However, what is attached to the other end of the station drop may be critical. If it is a workstation or terminal, the impact may be insignificant. But if a host or server is attached, the delay in restringing the cable may be intolerable. In this case, spare runs to areas around the server may be used. Move the server from one connection to another either physically or with a long extension cord. At the cross-connect point in the wiring closet, swap ports on a MAU or change patch cords to reflect the new location (make sure this is document-

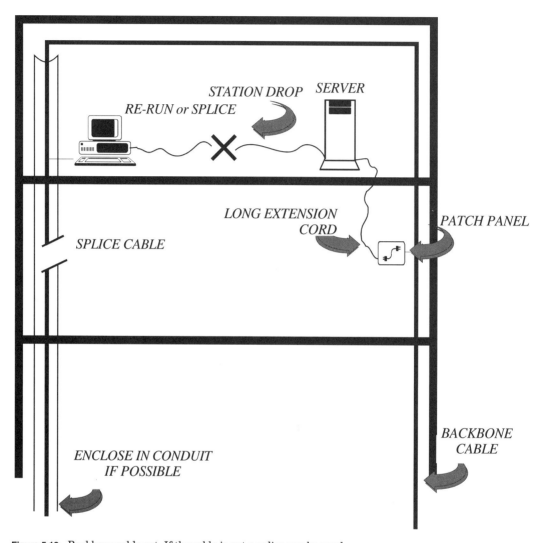

Figure 5.12   Backbone cable cut. If the cable is cut a splice can be used.

ed to the new configuration). This is contingent on the severity of the cut, and the reason and the actual number of station drops that have simultaneously been cut.

Wherever possible, the riser or backbone cables should be enclosed in conduit to protect them from cuts. This is encouraged for long-term reliability and protection. At the closets on each floor, the conduit will be open so that access to the backbone can be accomplished. This is the weak link in the network, since the cable is exposed to possible cuts and other risks (i.e., nicks, kinks, etc). In this closet it may be wise to add a couple of extra coils as shown in Fig. 5.13 for two reasons:

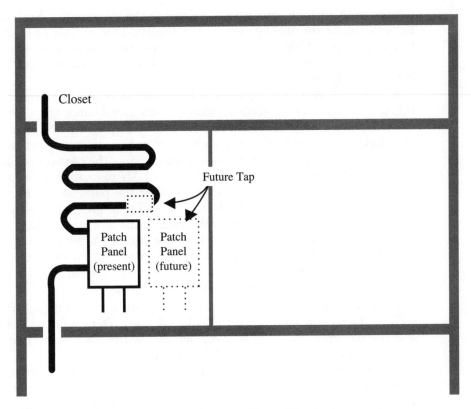

Figure 5.13   Using extra coils prevents future problems with extra taps or cuts.

1. The coils may be used to allow slack in case of snags. Furthermore, in the event of a cut, the coil allows for extra cable to put the necessary splice in place.

2. The added cable can allow for future taps or growth onto the backbone, depending on the cable systems used. If coaxial (i.e., Ethernet) cable is installed, this added coil allows for transceiver taps. If fiber is used, the coil allows for the necessary fusion of the fibers without rerunning the cable or creating added loss due to excessive splicing.

Each of these reasons merely adds to the depth of the cabling infrastructure to supply the capacities that may become needed at some future point in time.

Fire damage

This is alarming, since it implies a life-threatening situation (fire) has occurred. If fire burns the cable, it will be best to replace it totally. However, the replacement of cable may be delayed while other reconstruction efforts take place. Therefore some possibilities exist; these include the following.

1. Cross-connect from another floor (or LAN) to get critical servers and workstations back on line. This would require that spare cables on the new

floor be used and run from a wiring closet to the workstation. Such a line can be temporarily run through ceilings, along the base of floor or wall lines, or through conduit. Cables can be taped to floors or walls on a temporary basis, but should be kept out of main traffic areas, so that carts and foot traffic are not constantly on top of the cable.

2. Rerun all new cabling. This is both expensive and time consuming. At this point the use of new conduit may be considered, since the older conduits may be stuffed and full of burned cables. To attempt a rewire in conduits that have some bad cables and some good ones may cause further damage and disruptions. Use common sense when deciding the next course of action.

3. Relocate the entire department on the LAN to another floor or building on a temporary basis. At this point consider the possible use of a wireless network mentioned above. While this is still emerging, the use of wireless radio or infrared light may meet the need until more permanent plans are put in place. These wireless networks are fairly easy to set up and get around the lengthy lead times to set up a wired environment. Furthermore, to wire a temporary location creates a "sunk" cost for the wiring that will be left behind when operations return to the original floor.

In the situation of a wireless network, there are various options. These include the technologies listed in Table 5.3 below. Depending on the topology and the nature of the LAN services, as well as the costs associated with the quick setup needs, the choices still run the gamut.

Graphical representations of each of these are shown in the following figures. In Fig. 5.14 is shown the radio-based Motorola Altair system operating in the 18-GHz frequency range (microwave). The NCR Wavelan in Fig. 5.15 uses a spread spectrum in the 902- to 928-MHz frequency range, an unlicensed frequency band. Figure 5.16 shows the spread spectrum in the 902- to 928-MHz frequency range offered by Telesystems. Figure 5.17 depicts the infrared wave that operates in an invisible point-to-point light source by BICC, and finally in Fig. 5.18 is the infrared light diffusion technique offered by Photonics, Inc.

Each of these systems are quickly set up. In general, follow the steps listed in Fig. 5.19.

The wireless solutions serve the needs for departmental LANs nicely. However, the wireless segments may have to be connected to a backbone

TABLE 5.3   Options in Using Wireless LANs

Technique	Topology	Product
Radio (microwave)	Bus	Motorola Altair
Radio spread spectrum	Token ring	NCR Wavelan
	Token ring or bus	Telesystems–Arlan
Infrared point-to-point	Ring	BICC
Infrared diffused	Bus	Photonics

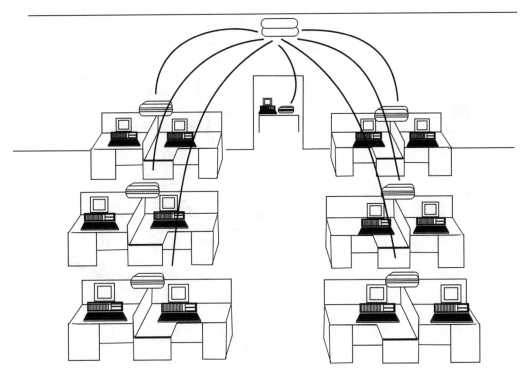

Figure 5.14 Typical Altair microcell.

cable. This can be accomplished by most of the systems on the market through various interfaces such as standard AUI, BNC, or TNC connectors. AUI denotes attachment unit interface and is a 15-pin connector. Thin Ethernet uses BNC, a "twist-on" coaxial connector. BNC actually stands for *bayonet Neill-Concelman,* but most people call it a bayonet network connector. TNC is for thick coaxial and broadband use, a "screw-on" connector. TNC stands for *twist Neill-Concelman*; most people refer to it as a twist-on network connector. Each scenario is different, so preplanned use and understanding of the selected system are prerequisites for use.

In Fig. 5.20 a wireless connection is used to provide workgroup services onto a backbone cable.

When dealing with cable damage caused by fire, the planning and recovery process must take other conditions into account. An example of this is to consider who is in control of the situation.

Assume that a fire has occurred and that the recovery process includes setting up a recabling or a wireless LAN for temporary service. The primary concern is to be assured that access to the office space is available. In many cases the fire department may deny access to the entire building until an investigation of the cause if conducted. The building inspector may control the building until a safe reoccupancy certificate can be issued,

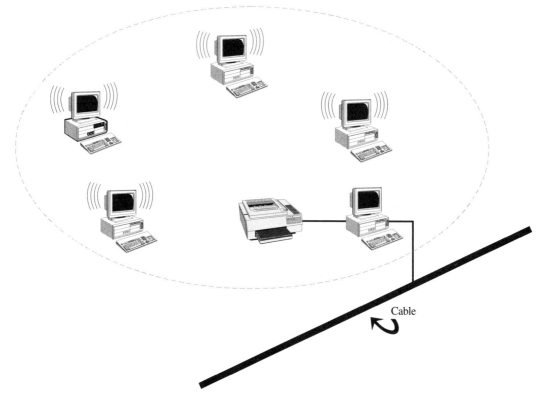

Figure 5.15   A spread spectrum radio system, i.e., NCR Wavelan.

indicating that other risks are not evident. Security may also be in control to guard the area until basic security, fire and safety tools, protection, etc., are reinstated.

All plans to recover the network in place may fall by the wayside if access to the space cannot be obtained. Therefore it is important to have a secondary plan ready to recover elsewhere and get back into business quickly.

Water damage

Once again, the incident that caused the water damage affects the decision process. Is the water from flooding on lower floors? Has the damage occurred from a sprinkler head (or other overhead conditions such as an overflowing sink or commode), or has the damage resulted from fire department personnel spraying the area during a fire?

Flooding would imply that not only is the cable going to be wet, but everything around it will be too. Cables must be dried out or else the protective sheathing will mold and/or decay. Furthermore, cross-connects must all be

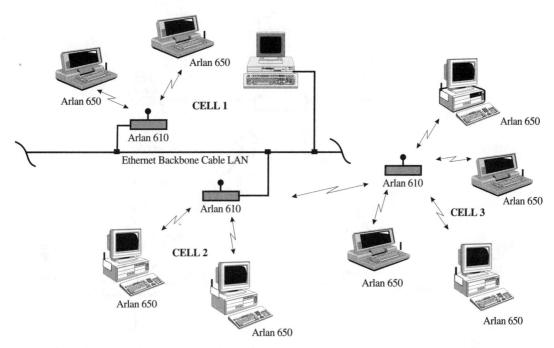

Figure 5.16 A spread spectrum radio system by Telesystems, Inc.

removed, cleaned off, and repunched. The 66 blocks* or patch panels may have to be replaced. Electrolysis and corrosion will begin immediately, causing static, poor connections, LAN disruptions, and intermittent problems. Do not waste time: if the damage is extensive, replace everything.

On electrical connections, arcing is also possible; this can cause fires and/or other significant damage. Therefore follow the steps listed in Fig. 5.21.

Rodent damage

Hopefully this is not a big problem for your LAN. However, rodents do reside in buildings and seek out any food source they can find. They will gnaw on any cables in a building, causing open circuits and possibly short circuits on wiring. To prevent this type of damage you can use the following.

1. Arsenic-treated cabling in areas where the risk exists.

2. Steel-jacketed cables, although this is expensive and makes it more cumbersome to use the cabling.

3. Traps and bait traps to eliminate the problem, using a qualified exterminator.

---

*66 blocks are wiring blocks that are used to "punch down" the wires. These are used in a lot of installations. The "66 block" was used by the telephone company to connect 50 wires (25 pairs) to a closet, then cross-connect to 25 individual pairs to the telephone set.

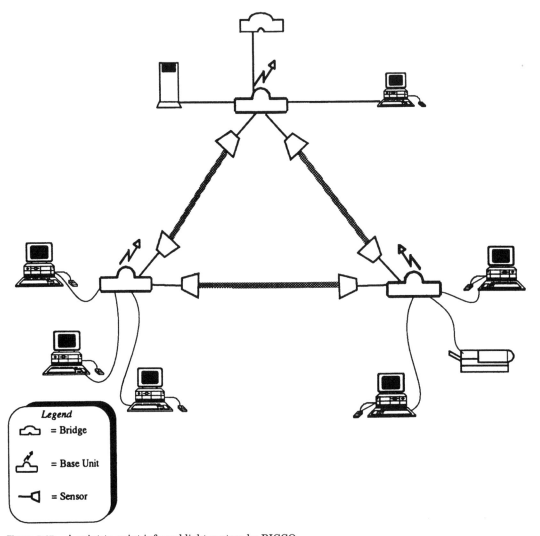

Figure 5.17   A point-to-point infrared light system by BICCO.

4. Use orange- or yellow-jacketed cables, which the rodents appear not to like. Many tests have proven that damage to yellow and orange cables is minimal compared to black, brown, blue, gray, and beige cables.

If damage occurs, new cable will have to be run since the protective coating will be lost, exposing the cable to greater risks and EMI and/or RFI interfer-

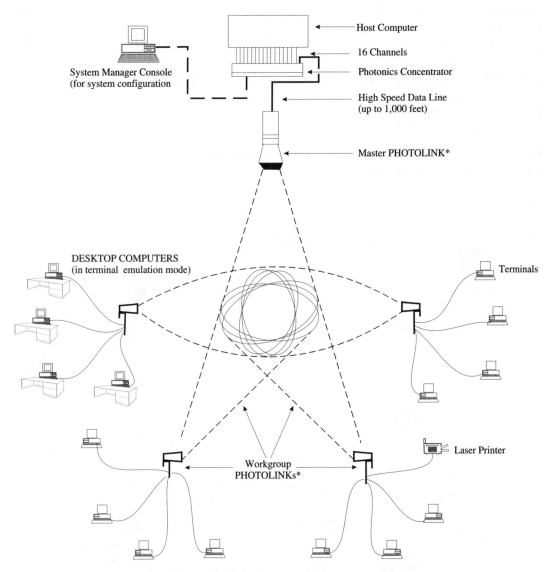

Host Computer

16 Channels

System Manager Console
(for system configuration

Photonics Concentrator

High Speed Data Line
(up to 1,000 feet)

Master PHOTOLINK*

DESKTOP COMPUTERS
(in terminal emulation mode)

Terminals

Workgroup
PHOTOLINKs*

Laser Printer

*Master and Workgroup PHOTOLINKS are interchangeable.

Figure 5.18   A diffused infrared light system by Photonics, Inc.

ence. If new cables are run, steps must be taken to eliminate the rodent problem. If preventative steps are not taken the problem will likely recur.

Other risks in certain parts of the country arise from insects (such as fire
ants). Use a certified exterminator to overcome and eliminate the problem
with these insects.

Whether infested by rodents or insects, damage can occur in ceilings,
under raised floors, in walls, and in raceways or conduits. Treat these cables
in advance.

1.   Open the PC.
2.   Install the wireless card.
3.   Close the PC.
4.   Install point-to-point nodes using line-of-sight alignment, control modules, or diffusion reflectors.
5.   Load the drivers.
6.   Add addresses to network.
7.   Attach physical connectors as necessary.
8.   Run the LAN.

Figure 5.19   Steps to install wireless LANs.

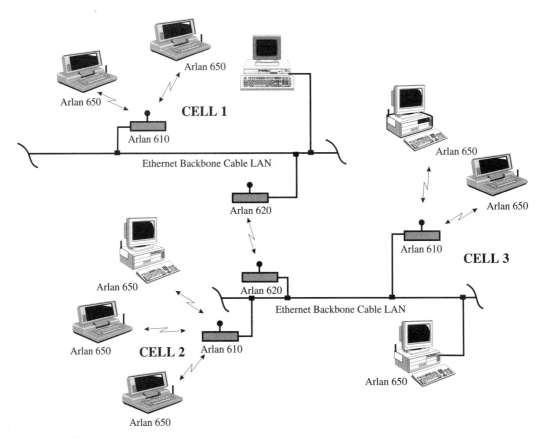

Figure 5.20   Linking radio-based systems to backbone cable systems.

EMI and RFI

If electromagnetic interference (EMI) and radio frequency interference (RFI) are evident, they will cause noise on the LAN. This will cause distortion to your data, creating errors. Consequently measures must be taken to protect the cable from these interferences.

1. Disconnect all electrical power and all devices. This should be done when electrical shock is not a threat. Use qualified trained personnel only.
2. Begin dehumidifying immediately to minimize the humidity and corrosive effects as soon as possible.
3. Use dryers or blowers to move the humidity out of the area.
4. Clean all contacts with a wire brush.
5. Remove contaminants from connectors with a wire brush and other solutions of deionized water or chemicals.
6. Replace any destroyed equipment.
7. Reseat all connections.
8. Repower the system, after all water and humidity have safely been removed. Have a qualified and trained electrician do this.
9. Never take chances with electricity—severe damage and threats to life are significant risks.
10. Rerun all diagnostics.
11. Check the cable with a time domain reflectometer for open and short circuits.
12. Verify that the cabling is ready for use.
13. Reattach devices (MAUs, workstations, etc.).
14. Bring services up logically in a low-risk sequence.
15. Document the new connections, equipment, or any other changes.

Figure 5.21 Sequential recovery procedures for water-damaged cables.

1. Find the source of the EMI or RFI and either move the cable or remove the EMI generation.
2. Check the cable for open circuits or ground loops, which could cause the cable to act as an antenna.
3. Shield the cable in the conduit; use shielding or a copper jacket to carry EMI or RFI away from the cable.
4. Check the quality of the cable, cable distances, and speeds of data to ensure you are conforming with specifications and standards. Use only high-quality cables and maintain the proper distances from electrical sources per manufacturer's specifications.

These points are summarized as a checklist in Fig. 5.22.

Media Access Units

Access to the cable system and the LAN backbone is through a media access unit (also called a multistation access unit). These devices provide for the connection of multiple stations or servers to the LAN. Electronics are used to provide the electrical and mechanical interfaces of the devices to the cable. From time to time, things can go wrong with this hardware. Closets could be very tight for space; yet multiple MAUs may be installed in these closets. Workers performing the installation could create many problems while at-

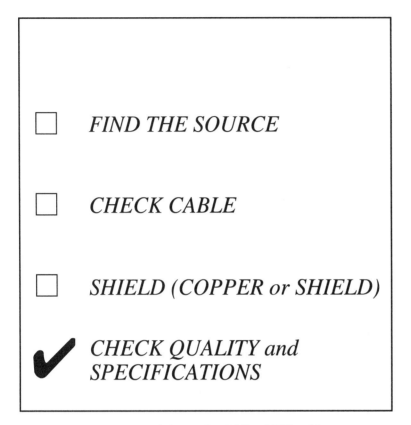

Figure 5.22   These points can help to isolate EMI and RFI problems.

tempting the installation. The environment must foster openness and comfort that allows the installer to be cognizant of the impact that dropped equipment will have on the network.

Droppage

If concern exists for liability of the damaged unit, the installer (whether a contractor or an employee) may feel compelled to install a dropped unit. A policy detailing who is liable and how the issue will be handled should be put in place. Short-term solutions may lead to long-term problems, so a sense of reality must be applied when developing the policy. Use Fig. 5.23 as a guideline for handling dropped units.

Electric spikes

Many problems have been caused by surges and spikes. These have become a way of life in business and industry. A MAU that is hit with a spike of elec-

- If a unit is dropped you should replace it immediately. Do not worry about whether it is good or not. The unit can be tested later by a qualified repair shop or returned to the manufacturer for replacement.

- Do not assume that it did not fall far! How far is safe versus how far is not can cause much longer-term problems. The key here is to be aware of intermittent problems that could occur over time if damaged MAUs are installed on the network. The best rule to use is replace it now and figure out if it is still good later.

- Spare units should be kept on hand to replace damaged ones. It is a fact of life that humans err. Spare units (numbering 5 to 10 percent of those in use) should be on hand for immediate replacement. Relying on a vendor to have spare units may cause unnecessary delays in replacing faulty units quickly.

- Replacing the unit will prevent future problems of an intermittent nature. The long-term impact could result in equipment failures, damaged or corrupted data streams, and finally network downtime. Enough risks threaten the LAN without adding to this problem. To paraphrase an old cliche, an ounce of prevention prevents a pound of trouble at a later date.

Figure 5.23 What to do if MAUs are dropped.

tricity can suffer either immediate or future problems. The MAU is designed to provide electrical isolation for the workstation so that a spike will not run right down the wire to the workstation and destroy it. To provide this isolation, breakers will quickly shunt the electrical current from going through to the drop cable. Furthermore, the MAU provides a form of digital repeating or regeneration to keep the appropriate voltage level on the signal that should be present. If these MAUs are hit with spikes, they could shut down immediately, causing disruption to the devices on it or shutting down the continuous flow of the signal. Either way, problems with downtime arise. A longer-term impact would be the result of spike or surge that degrades the MAU's electric capacity but does not totally destroy it. Once again, the risks are there but may not be totally evident.

Any piece of electrical equipment should be protected from spikes and surges. This means that a surge protector can be placed between the power source and the MAU. Should the spike be stronger or faster than the protector can handle, the spike will leap across the protection and damage the equipment. Never assume that the surge protector will prevent damage all the time. However, do not ignore its benefit for day to day glitches.

The checksheet shown in Fig. 5.24 should be used if a MAU is hit with a spike, or if power is suspected of causing damages.

Overheating

In a data processing environment, a good deal of attention is paid to environmental conditions. The primary areas covered are temperature and humidity control for efficient equipment operation. The data processing shops will look at the range of heat and humidity and stick to the manufacturer's specifications for operation. However, in a LAN environment, many of these issues get either pushed aside or ignored. LAN server and equipment rooms are not predefined areas nor are they treated for special operations. Moreover, they

Replace the unit immediately.

Remove the source of power. Disconnect the power and have the source checked by a qualified electrician.

Install a surge protector.

Install a new media access unit.

Power the unit up and run diagnostics and the self-test. Verify it is operating correctly.

After testing, reconnect the station units (i.e., PCs, printers, workstations) and the connection to the LAN (bus or ring).

Figure 5.24 Checklist of steps to follow if a MAU is hit with an electric spike.

are often the first empty space that can be commandeered. Generally the closet that serves as an electrical room, telephone wiring closet, and storage (for the janitor's deep sink) area will be selected as the LAN equipment room. This is usually due to tight space where these closets are already set aside.

Packing servers, hubs, concentrators, and/or MAUs into a closet of this type always begins a process that poses risks. As more and more equipment is added and as security of the room dictates keeping the door closed, heat will inevitably begin to build up. Coupled with heat buildup, the air is kept very still, creating a dense heat pocket (or oven effect). The best way to solve this problem is to obtain a flow of air or a draft to move the air through the room and vent out the heat. This can be done by using the information listed in Fig. 5.25, which is a physical facilities list of procedures. Obviously the best solution is to install the electronics in an area that is environmentally protected against heat. Reality, however, dictates that this service cannot always be provided if the LAN is kept inexpensive. Therefore, the second choice is to work around the problem by moving the heated air out of the space. This situation is represented in Fig. 5.26, which allows for inexpensive room treatment using vents in the door and walls, a heat sensor set at 85°F temperature to alert the system manager, and a ceiling fan that will move a minimum of 500 ft³/min [cubic feet per minute (CFM)]. This should provide enough movement to prevent heat buildup in the room.

1. Place fans in the room to move air around the area.
2. Cut vents (if possible) at 8-ft-high intervals to move heat out of room.
3. Use an indoor vent to create a draft flow from outside the door through the room to the vent.
4. Move air at manufacturer's specifications to keep the room at the desired temperature.
5. If the units are intermittently failing, replace them with spares and have the old ones tested and repaired.
6. Place a heat sensor in the room with alarms wired to system managers.

Figure 5.25 Physical facilities list of procedures for avoiding heat buildup.

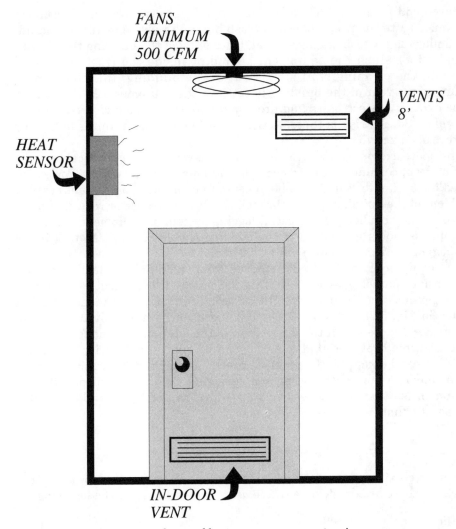

Figure 5.26 Inexpensive vents, fans, and heat sensors can prevent major buildup of heat.

Loss of units

When the issue of lost or stolen MAUs comes up, many users say it is impossible in their environment. Certainly this is an extreme case, but who knows what ultimately may happen in a building?

An aftermarket for MAUs has not yet emerged, thus the resale of these devices seems far fetched. Stranger things have happened in the LAN marketplace from unlicensed copies of network software, viruses being introduced from the least likely source, and loss of major amounts of equipment. Security is always an area in which users feel they are well protected. However, a case in point deals with a cleaning employee who removed equipment in a client's office area. The cleaner would disconnect equipment after hours, hide it in a trash bin, and remove it from the building under the guise of trash removal. Once outside the building the cleaner would either stash the equipment behind or inside the trash dumpster. After all the cleaning staff signed out of the building, this thief would return to the rear of the building to retrieve the equipment. MAUs appear to be expensive electronics and are usually kept in locked closets, indicating value. So, the risk is there. Actions that follow after a unit is lost is the more crucial activity.

The above example may seem unreal and may even be impractical to some. But other opportunities to remove this equipment are available. For example, a technician (or someone who appears to be a technician) could possibly walk out of the building, either through the front entrance or rear dock area, without being challenged. A special logging system of equipment (such as a parcel pass) going into or out of the building should be used. A quick list of this logging activity is shown in Fig. 5.27 below.

Any equipment that is to be removed from the building should have an appropriate parcel pass with it. This pass should be signed by an authorized person (i.e., security, facilities, administrative manager, etc.), as shown in Fig. 5.28. This pass should be used for both employees and non-employees, coupled with the disposition of the equipment. The form should have multiple copies so that one copy can be kept in a security file, a second copy can be kept by the security guard or reception desk person, and a third copy can stay with the person removing the equipment or with the equipment so that if it is returned an audit trail of log-in and -out dates can be established.

Servers

Servers are extremely critical assets on your LAN, regardless of their true function. These are the devices where users access and store the following:

1. Files

2. Applications

3. Print queues

4. Directories

---

1. Although unlikely, if units are lost or stolen, replace them immediately.
2. Check logs on who had access to the rooms.
3. Verify repair logs to see if the units were removed for maintenance.
4. Secure the room—change locks if necessary.
5. Notify security.

Figure 5.27 List of logging activities.

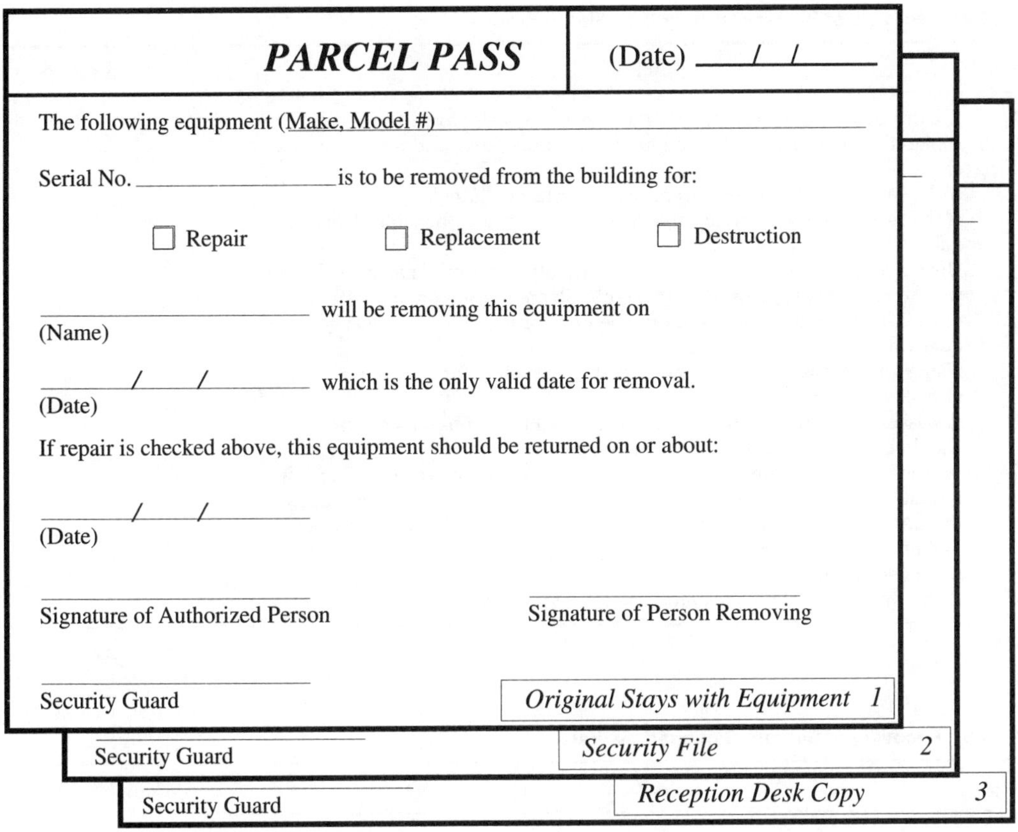

PARCEL PASS	(Date) ___/___/___

The following equipment (<u>Make, Model #</u>) _____

Serial No. _____ is to be removed from the building for:

☐ Repair   ☐ Replacement   ☐ Destruction

_____ will be removing this equipment on
(Name)

___/___/___ which is the only valid date for removal.
(Date)

If repair is checked above, this equipment should be returned on or about:

___/___/___
(Date)

_____
Signature of Authorized Person   Signature of Person Removing

_____
Security Guard | *Original Stays with Equipment*   *1* |

Security Guard | *Security File*   *2* |

Security Guard | *Reception Desk Copy*   *3* |

Figure 5.28 A parcel pass for removal of equipment.

5. Communications

6. Other

Therefore, if a failure occurs, the server will have to be brought up prior to the individual user workstations. Particular attention will have to be apportioned to ensuring that the servers are functioning properly. Only after these servers are running properly should critical users be added to the network.

Table 5.4 is a step-by-step procedure to recover servers from damage caused by fire or smoke. Contaminants on the server or cards can cause extensive damage. Treat these quickly and efficiently. Figure 5.29 is a flow-chart designed to assist in the recovery of servers, regardless of the problem. This flowchart walks a user (or vendor) through the process easily.

Table 5.5 is similar but deals with water damage, through which heat inside the server builds up. If cold water is dumped on the server, the cards could crack from the sudden change in temperature. Visible inspection is necessary; humidity must be removed as soon as possible. A heater may be used on low power to dry the equipment quickly.

TABLE 5.4   Recovery of Servers Damaged by Fire or Smoke

Step	Action	Completed by
1	If a fire has occurred it is likely that the server will have to be restored or replaced. Disconnect power and network connections, and move the server out of the area.	
2	If the server is totally destroyed, replace it immediately.	_____
3	If the server appears salvageable, remove the cover. Check for debris in and around the bus and cards.	_____
4	Have a qualified repair technician remove all cards (expansion cards, network interface card, motherboard, etc.). Clean all components with a deionized solution (water) and dry the unit completely. If necessary relacquer cards.	_____
5	Using appropriate test equipment, test all circuitry for continuity, short and open circuits, etc.	_____
6	Reassemble the unit. Power up the server and run diagnostics. This should be done with a bootable floppy disk.	_____
7	Access the storage subsystems to ensure everything is working properly.	_____
8	Reinstall the unit onto the network and attempt to access the server from the LAN. Be sure that its address is recognized and that access works, and test for any errors.	_____
9	Let a user terminal access the server. Verify operation.	_____
10	Begin migrating users back onto the LAN and servers.	_____

TABLE 5.5   Recovery of Servers Damaged by Water

Step	Action	Completed by
1	Cut off power if it is safe to do so. If in doubt, contact appropriate maintenance personnel.	_____
2	Move the unit to a safe and dry place.	_____
3	Open the cover and allow the unit to dry out thoroughly.	_____
4	Test all cards and remove all contaminants (rust, minerals, oxidized residue, etc.). Electrolysis will begin immediately.	_____
5	Have a qualified technician remove all cards and deionize them. Check all circuitry with appropriate test equipment. Redip or relacquer cards as necessary.	_____
6	Reassemble the unit and power it up. Boot from a floppy disk to test the components. Run diagnostics on the system. Replace any failed cards.	_____
7	Add drivers, cards, and other devices in a logical sequence. Test each one as they are added to the system.	_____
8	Reconnect to the LAN and check the system, addressing its units, etc., thoroughly.	_____
9	Add a user device to access the server.	_____
10	Verify operation and add other users.	_____
11	Check everything for intermittent failures or glitches.	_____

Physical damage

The easiest way to prevent servers from being bumped or knocked over is to lock them in a room and limit access. However, the risk still exists that such

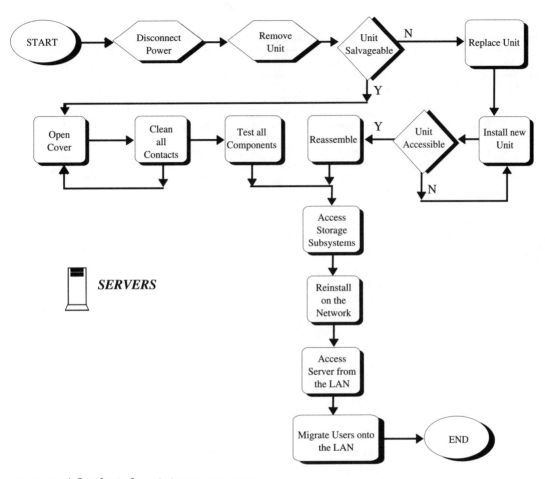

Figure 5.29 A flowchart of events to recover servers.

an event will occur. Table 5.6 summarizes the actions required in the event a
server is bumped or knocked over.

Inaccessibility of Building

In the event of a natural disaster (earthquake, flooding, hurricane, etc.) or a
human error that denies access to the building, recovery to a remote site
becomes necessary. Have all components necessary to reestablish critical services, in advance, ready to use.

Table 5.7 summarizes the steps required to get back into business even
though the building is inaccessible. This of course assumes that the data and
documentation is stored off-site.

Figure 5.30 is a graphic representation of the recovery process at a new
site. In this scenario, the backup tape (or other media) *and* documentation

TABLE 5.6   Recovery of Servers Damaged from Being Bumped or Knocked Over

Step	Action	Completed by
1	Cut the unit power off. Disconnect it from the LAN.	_____
2	Check for any visible damage. Repair as necessary.	_____
3	Secure the components; secure the server to a stable surface.	_____
4	Power the unit up and run diagnostics. Run "chkdsk," etc. Verify proper operation; replace any failed components.	_____
5	Reconnect the server to the network. Verify access and operation.	_____
6	Secure the server and lock the room. Limit access.	_____

TABLE 5.7   Checklist for Recovery if the Building is Inaccessible

Step	Action	Completed by
1	Recover files and applications from tapes or other media and load onto a new server at the recovery site. Test all systems before going on-line. Use the documentation on hand to reestablish the network.	_____
2	For servers with remote access, log on from the new LAN and download your files to the new LAN. Prior to doing this, ensure security is maintained. Make sure modems can be turned off from the remote site.	_____
3	As users create new files, or access old ones, ensure appropriate backups or incremental backups are created for later restoral onto the original LAN, when you move back to the original site.	_____
4	Secure the old building to ensure that theft or unauthorized access is prevented. Tighten security as much as reasonably possible.	_____

are kept off-site, and both are retrieved from the storage site and delivered to the recovery location.

Terminals, PCs, and Workstations

All of these units are designed around the end-user device. Each of them is exposed to the same risks. Such risks include the following.

1. Physical damage from being knocked over or dropped

2. Spillage

3. Flood and fire damage

Regardless of the risk or the event, these units are the simplest to recover. Follow the steps in Table 5.8 to recover individual devices.

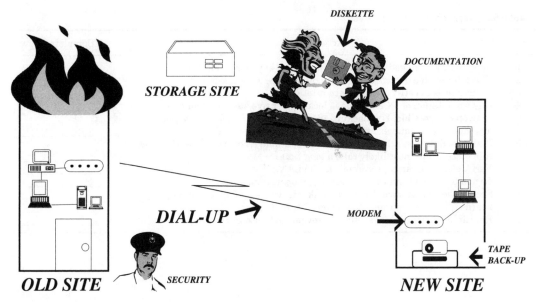

Figure 5.30   Retrieve backup tapes and documentation from the storage site.

TABLE 5.8   Terminal and PC Recovery

Step	Action	Completed by
1	Replace any damaged components (monitors, keyboards, etc.).	_____
2	Check for visible defects or damage and replace as necessary.	_____
3	If the unit is questionable, replace it totally.	_____
4	Any workstation or PCs that have local hard drive storage should be recovered if possible. Use backups to recover data. If backups are not available, use whatever utilities are available to recover data.	_____
5	If all else fails, send the hard disk to a reputable recovery house. However, this process will most likely take a long time.	_____
6	Verify clean operation, reconnect to the LAN.	_____
7	Log on to the LAN and verify access and operation.	
8	Secure the unit to a solid surface, desk, etc. Use a spill shield on keyboards (plastic membrane) to prevent future damage.	_____
9	Retrain users on the risks, etc., to prevent further occurrences.	_____

6

Departmental Recovery

When considering departmental recovery, options exist to use various techniques in recovering from a disaster. The software that must be reloaded depends on the nature of the disaster. If a total failure occurs, everything must be recovered and loaded. This includes the following:

1. Network operating system software

2. Application software

3. Individual files

4. Files from tape

5. Files and applications that have been archived

Typically, a network crash will only be involved with current files, the recovery of which is a little easier to perform, as long as the system is running and the network applications are intact.

Backup Systems

Traditional methods of recovery include backing up files that are under end-user control. It is hoped that floppy disk backup or tape backup systems have been used. These are good approaches for stand-alone users, but when a network is involved, the intricacies are different. Moreover, individual users become lax in their efforts to back up their data. They expect the LAN manager to do it for them, especially if they know this can be done.

In Fig. 6.1 the flow of how the overall backups work is shown. In this scenario individual users who must create floppy disks may start out religiously doing backups when the system is new. However, over time, as the files grow larger, more floppy disks are needed and the time necessary to create a backup is extended. Thus, it becomes a nuisance. Slowly, users begin to let more time slide between backups until these become virtually nonexistent. This is

Figure 6.1   Using floppy-net erodes backup processes quickly.

normal. As long as the process is prolonged it will erode. So plans must be put in place to accommodate the backup process.

Therefore, many companies have hired part-time employees whose responsibilities include a weekly backup of individual PCs or servers via floppy disks. They produce the disks and label and store them. However, as time goes by even this systems erodes. Records get sloppy; labeling becomes abbreviated, and turnover of these human resources becomes an administrative problem. People leave, new ones are hired. The new person's training becomes less intense and problems perpetuate. Another problem with this method is that the part-timer evolves into a person of all trades. Slowly but surely this individual becomes more visible within the organization. Department managers begin to sense that backing up systems is not a full-time activity, but they have tasks that this part-timer could do in the interim. Additional duties get piled on the designated backup person until the real job for which this person was hired becomes secondary in nature. Since the backup process can be tedious, this person begins to migrate to more enviable or enjoyable tasks. Thus, the system of backing up the PCs or servers breaks down rather quickly. Shortcuts are taken to get through the mundane tasks; backups become further apart, until finally they just are not done anymore. Unfortunately, this may sound comical, but the ramifications are severe. When (not if) the server or an individual PC finally crashes, the last backup may be totally useless.

The part-timer then leaves the organization, disappointing the staff. This person did not do the job that he or she was hired to do. Consequently, replacing this individual is not something that will be easily accomplished. Everyone will only remember the negative side of this position. But what really caused the breakdown in the process? It was the departmental manager's lack of support for the process that led to this shift in responsibility or loyalty! If managers do not perceive the need for backup, the employees will also adopt this philosophy. What happens after an incident of this type is that the backup just does not get done. Another possible outcome would be to start the process all over again. However, reality sets in quickly and the part-time position becomes a less desirable option.

To overcome the above-noted problem, a potential area that can be tapped for assistance is the data processing shop. A transition from a part-time employee to a data processing shift operator is an easy hurdle. Many data processing managers are fighting to gain back control over these information resources. Therefore they will go to great lengths to take back what they feel that they should never have lost. They will volunteer to do the LAN backups using existing staff. This will show the organization that the data processing staff is still committed to information resources.

With volunteering to perform the backups by the data processing staff, a different perspective creeps in. Ownership of the responsibility may pass hands. But the issue of ownership and control of the data may become a point of contention. The real questions are who owns the data and who is responsible for it? The means of addressing the questions have been handled in varying ways by different organizations, as typified below.

1. Some organizations feel that the data belongs to the business (corporation, firm, etc.) and therefore it is the business's responsibility to secure it, back it up, and control it. The responsibilities therefore are doled out to a single entity or department such as data processing.

2. Other organizations feel that the data belongs to the individual division or department that creates it. Therefore data integrity, backup, and control remain with the division or department.

3. Lastly, some organizations have taken a position that the data belongs to the individual, and therefore the individual should provide the necessary backup, integrity, and security.

Although all three scenarios based on management's decision and direction work, the risk of loss is still high when choices 2 or 3 above are selected. A possible solution to this problem might be that files, applications, etc., that reside on the servers belong to the organization and will be appropriately protected, whereas files, directories, and so forth that reside at the individual workstation are the responsibility of the individual user. In two of these scenarios the data processing manager can serve as the agent responsible for the organizational or divisional data. This is why the offer from data processing is made. Unfortunately, what often happens is that the second or third shift operators become totally responsible for the necessary backups and security, which could create a separate issue.

In these recessionary times, the third shift may be the first employees laid off to save expenses. Thus this shift may become nonexistent, leaving a gap in the backup process. A hit-or-miss approach results, which invites disaster and exposes the data processing function to the feelings of old by the users. (Recall Chap. 1 where the users wanted to take control because data processing did not do the job the way users wanted.) Figure 6.2 reflects this process of the shift in responsibilities for tape backup systems.

Tape backup systems were introduced at the individual PC level to aid in the process. Instead of using tens of floppy disks, a simplified tape system allows for quick and convenient backups. Again, however, the process erodes from once a week, to once a month, to never. The net result is that too little emphasis is placed on backups in some organizations.

Example: An organization that used a stand-alone PC installed a tape backup system to provide coverage. The backup was the responsibility of the user, which worked fine for the first few months. When a disk crash occurred several months later, the user was embarrassed to report that the last backup was three months old and that the data was useless. After much rekeying and anguish on everyone's part the data processing department installed an automated procedure and tape system that would back up the system on a regular basis (weekly). This was fine but the procedure required that the

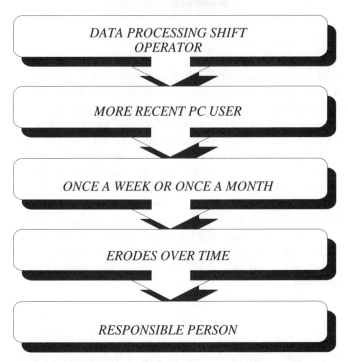

Figure 6.2   The process of tape backups may shift from a data processing operator to a PC user.

users put the tape cartridge into the drive. You guessed it! They were not putting a tape into the system when the systems alerted them. When they came in on the following workday, they just canceled the message from the system alerting them that backup was required. A head crash occurred several months later, leaving them once again embarrassed to report that the most recent backup was too old to be of any use. Thereafter, a data processing person was assigned to perform the backup on Friday evenings after the user went home. *Whose responsibility is it?*

In a LAN environment with multiple users sharing the same file and application servers, this becomes even more critical. Recovery can be as simple as creating a daily or weekly backup tape on the servers and restoring from these tapes. Critical applications require more frequent incremental backups. This timetable can be established on an individual basis. A system or disk crash will be disruptive but the data can be recovered within a day (or a few hours), depending on the amount of data needed.

To overcome these risks many additional systems have been introduced in the LAN arena. These include both software and hardware solutions that can provide various forms of backup or redundant services; these are summarized in Fig. 6.3 and shown graphically in Fig. 6.4.

The complexities and costs of each of these systems varies by the system chosen. Examples of these systems follow.

Disk mirroring

Figure 6.5 depicts a single-access method of disk mirroring, which may be the first line of defense for protecting the data: a fault-tolerant method protects against hard disk failure by writing to disks simultaneously over the same channel. If one drive should fail, the other drive can keep the system operating without loss of data or interruption to the user. Obviously the ability to write to two disks gives users an immediate standby on the data that is written to disk. If a single disk failure occurs the second is ready to go. However, as fault tolerance goes this is a level above doing nothing at all, but other means can achieve better results. Make no mistake; any redundant service is at least a first step. Budgetary constraints or lack of management commitment may limit the options. Since data protection is a personal call, the protection of the LAN and related data cannot be overly expensive or management may perceive a negative image. Therefore the first level of defense may be disk mirroring.

Disk duplexing

After instituting a disk-mirroring approach, the next logical step would be to begin budgeting for channel and disk duplexing. In Fig. 6.6 duplexing shows how a second channel into a single server provides more redundancy. This scenario uses disk mirroring and two separate channels to write to these two disks. This provides a fault-tolerant method that protects against a hard disk or bus channel (card) failure between the file server and the disk. It is an extension of disk mirroring. This involves a redundant controller, power sup-

---

1. Disk mirroring (a redundant service)
2. Disk channel duplexing (a redundant service)
3. Tape backup
4. Optical backups
5. On-line systems shadowing (a redundant service)
6. Virtual disk systems (a redundant service)
7. RAID (redundant arrays and inexpensive disks) technology (a redundant service)

Figure 6.3 Redundant and backup service options.

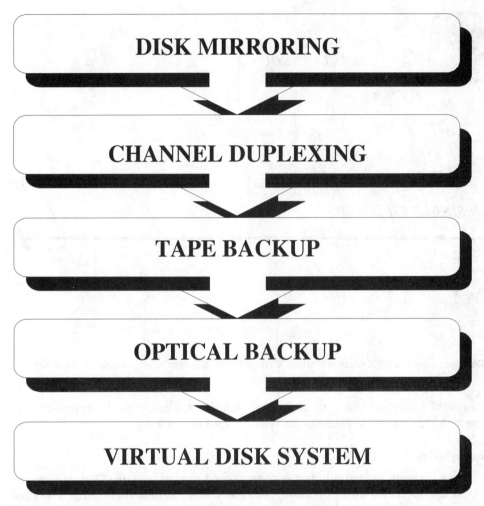

Figure 6.4 Various options are available for backup and redundancy.

ply cable, and hard disk. All writes are sent through both channels to the disks simultaneously. If a failure of one element occurs, the failing channel is shut off from data requests. The other disk on the separate channel can keep the system operating without disruption.

Ultimately the cost of both channel duplexing and disk mirroring is only marginally more expensive than disk mirroring alone, but the dual-redundancy process provides a lot more redundancy. A planned effort is needed to provide this level of tolerance. The next step to be considered would be the use of duplexing and mirroring to two separate servers, as shown in Fig. 6.7. Writing to two different drives served by two separate channels and connected to two servers offers the best of both worlds with triple redundancy to prevent a single failure from disrupting the access to the networked data. This

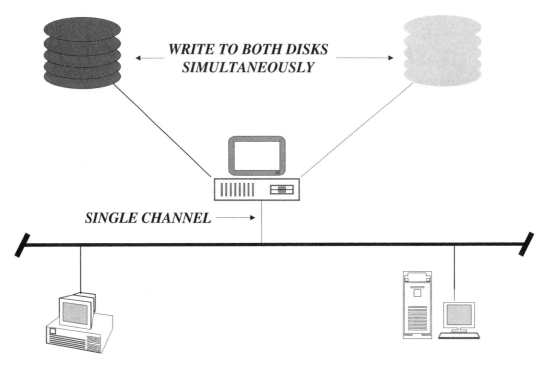

WRITE TO BOTH DISKS SIMULTANEOUSLY

SINGLE CHANNEL

Figure 6.5 Disk mirroring is a fault-tolerant method. Writing to two disks provides redundancy of the data.

implies two complete sets of data writes. It is a more costly approach but may be the ultimate solution for real-time access to data.

Tape backup systems

These systems are usually faster than other forms of backup. They are secure and can be configured to perform an unattended, automatic backup. The files are shown in Fig. 6.8, which describes the difference in the backup procedures.

Tape systems allow fast file restore capabilities to locate and move tape to a desired file. Using a tape backup system, file grooming (the purging of unused or rarely used files) can be accomplished. The systems may also provide a library or cataloging service so that files can be quickly located. Newer versions of these tape subsystems can accommodate dynamic log-on/log-off to the server by the backup process. These capabilities are shown in Fig. 6.9. The services of the tape system can accommodate unattended operation and servers. Cataloging is a newer feature. Using a standard tape rotation schedule as shown in Fig. 6.10, a user should be able to provide reasonable backups. Many larger LAN managers will conduct a full backup nightly.

A second version of the tape backup system is a dedicated tape server that is accessible to all users. This is shown in Fig. 6.11 and provides even better services on the LAN. A user can designate the backup of the files or critical needs without placing the burden on the LAN administrator. These

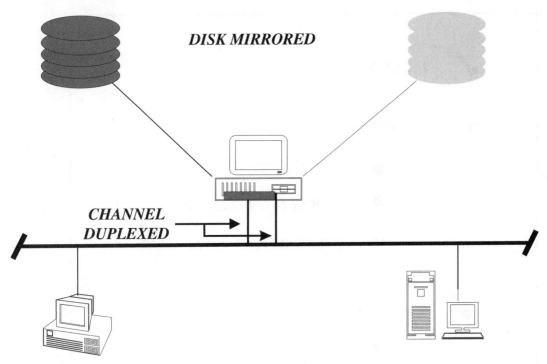

Figure 6.6   Disk duplexing uses two separate disks and channels as a better fault-tolerant method.

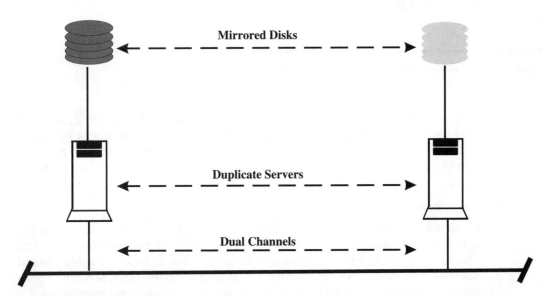

Figure 6.7   Using two different server channels and disks is better yet.

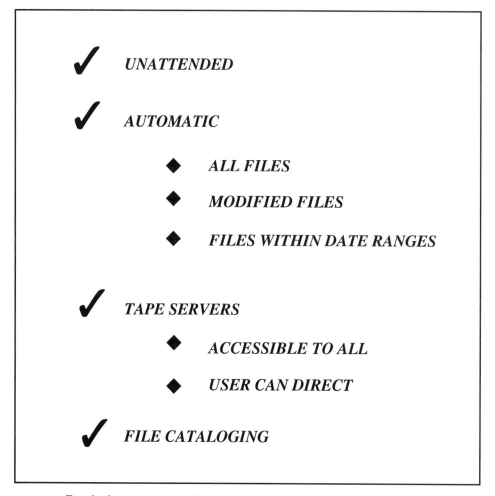

Figure 6.8 Tape backup systems provide these types of services.

systems also provide file cataloging, allowing users to retrieve files and data more easily. Maynard Electronics has a system that also provides a peer-to-peer backup capability, allowing a tape backup of a user's DOS system by the LAN administrator.

Optical backup systems

Different types of optical media provide different types of capabilities. In the optical world the primary means of backing up systems, applications, and files is through either write-once read-many (WORM) or erasable optical devices. The pricing on these various systems varies considerably. These devices differ from a tape backup machine since tape is a sequential service

FULL BACKUP	**SAVES ALL DATA, EVEN IF IT'S BEEN BACKED UP BEFORE.**
INCREMENTAL	**SAVES ALL DATA WHICH HAS CHANGED SINCE THE LAST BACKUP.**
DIFFERENTIAL	**SAVES ALL DATA SINCE THE LAST FULL BACKUP.**
SELECTIVE	**SAVES ALL DATA SELECTIVELY TAGGED BY THE END USER.**

Figure 6.9 Tape backup systems can provide full or partial backups, as shown here.

whereas optical medium is a random access unit, allowing direct access to a specific file. These systems can store large amounts of data for easy retrieval. A typical optical system on a network is reflected in Fig. 6.12. An example of a system using optical on-line storage systems would be LAN Sweep.[1] This is an optical storage service that is a hardware and software solution. Menu driven, the LAN Sweep product sits on the network and provides unattended backups of the servers (a typical configuration is one to four servers). A sample menu from a Novell Network using the optical backup is shown in Fig. 6.13.

---

[1]LAN Sweep™ is a registered trademark of WBS Associates, McLean, VA.

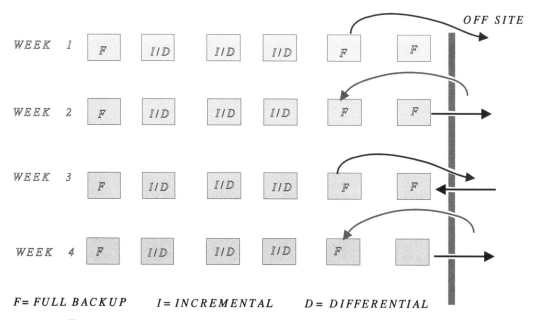

Figure 6.10 The rotation schedule should provide reasonable backups for the average LAN. (*Source: Maynard Electronics*)

This system can perform all variations of

1. Full backup

2. Incremental backup

3. Differential backup

and store them on the optical disk. Using compression techniques this device can store up to 1.7 to 1.8 Gbytes of information, as opposed to the typical 900 Mbytes. The disks are labeled using a DOS convention of up to 11 characters, with the label representing the time period covered.

The labeled disks can run (depending on the amount of data being backed up) several days to several weeks. When restoring data, simply specify the last date of the backup. The system will prompt the user for the appropriate disk.

Fully unattended operation is available to provide the backup. The LAN manager programs the time, date, and type of backup desired in a "profile." The system does the rest. In the event the optical disk is full, the system will alert the LAN administrator [or other designated person(s)]. When a new disk is installed, the system will carry out its procedure. By logging onto the LAN Sweep as a logical and a physical drive (e.g., D:), a directory prompt can be used to view the contents of the files by user, data, file, etc.

LAN Sweep also allows the use of some unique wildcards to locate files. The author uses RJB.<u>xxx</u> for files the author creates. The secretary uses SRJB.<u>xxx</u> for files created for the author. By using a *RJB* the system will

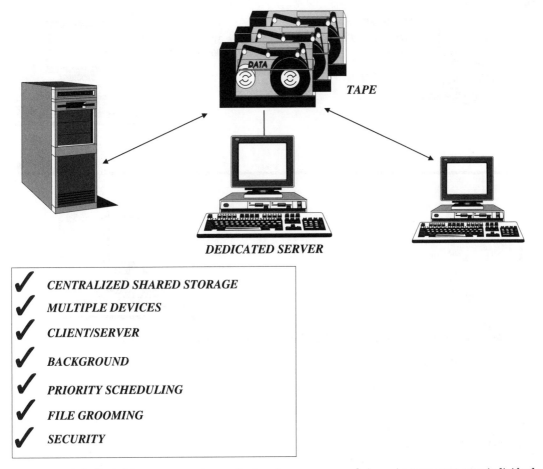

TAPE

DEDICATED SERVER

✔ *CENTRALIZED SHARED STORAGE*

✔ *MULTIPLE DEVICES*

✔ *CLIENT/SERVER*

✔ *BACKGROUND*

✔ *PRIORITY SCHEDULING*

✔ *FILE GROOMING*

✔ *SECURITY*

Figure 6.11 A dedicated tape server using a client and server approach is an improvement over individual tape systems. (*Source: Maynard Electronics*)

bring up all associated files with the specified three-character string anywhere in the file name. This makes it easier to find files when the file name convention is unclear or ambiguous. The screens for the restore process are shown in Fig. 6.14.

Virtual disk systems

Another system is the on-line or real-time continuous backup service. A system developed by Vortex Systems[2] is currently the only one such system on the market. The system uses fault tolerance and real-time backup service in a single product. It does all of the service functions as outlined in Fig. 6.15.

---

[2]Vortex Systems™ is a registered trademark of Vortex Systems, Inc., Pittsburgh, PA.

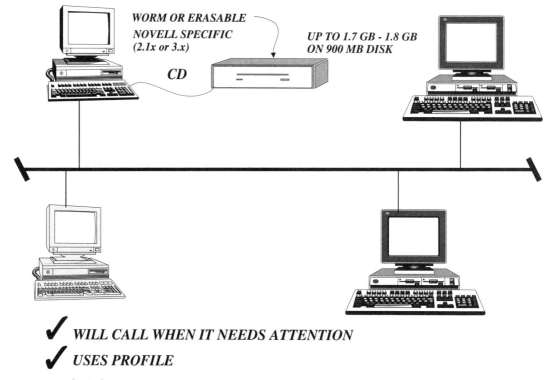

Figure 6.12   Optical systems can provide significant amounts of storage. (*Source: WBS and Associates, Inc.*)

1. BACKUP (FULL) MONTHLY
2. INCREMENTALS NIGHTLY
3. HUMAN-NEED ONE (LSHELP)
4. BACKUP IS PERMANENT

BACKUP   RESTORE   DOS   UTILITIES   HELP   QUIT

WAIT FOR MAIN EVENT		CURRENT PROFILE
INCREM. FULL DIFF. SELECT FILES SKIPPED FILES BINDERY		PROFILE
[RESUME/RESTART]		
ARCHIVE BIT RESET ON		

Figure 6.13   The main menu of the optical backup system. (*Source: WBS and Associates, Inc.*)

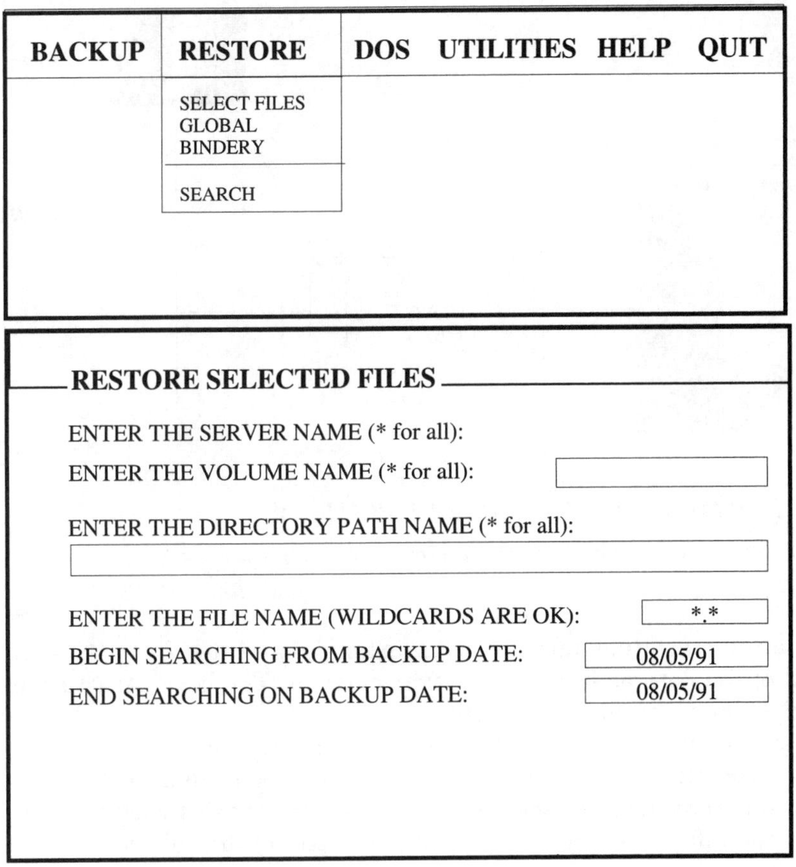

BACKUP	RESTORE	DOS	UTILITIES	HELP	QUIT
	SELECT FILES GLOBAL BINDERY				
	SEARCH				

RESTORE SELECTED FILES

ENTER THE SERVER NAME (* for all):

ENTER THE VOLUME NAME (* for all):

ENTER THE DIRECTORY PATH NAME (* for all):

ENTER THE FILE NAME (WILDCARDS ARE OK):   *.*

BEGIN SEARCHING FROM BACKUP DATE: 08/05/91

END SEARCHING ON BACKUP DATE: 08/05/91

Figure 6.14 Screens for the restore process.

1. Preserves data continuously
2. Provides "point-in-time" restoration
3. "On the fly" recovery without disruption
4. Isolates storage devices from the network
5. Allows "hot" removal and replacement of failed devices ("hot swapping")
6. Allows for event notification for immediate response

Figure 6.15 The service functions provided on the Vortex System.

Using the storage management system (SMS) for Novell, the Vortex system eliminates daily backups. Since it performs continuous backups both automatically and transparently, the daily backup really is not necessary. Provisions also exist to back up open files regardless of the file structure used. In the event of a crash the LAN manager merely turns back the clock

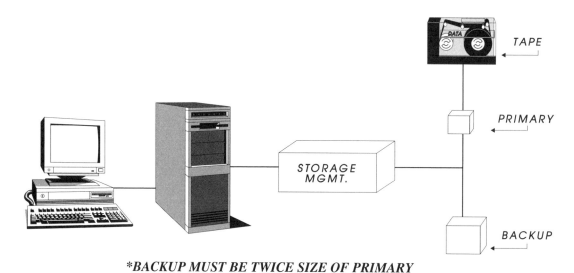

BACKUP MUST BE TWICE SIZE OF PRIMARY

Figure 6.16   The basic structure of the Vortex combined service. (*Source: Vortex Systems, Inc.*)

to just before the crash (one minute, as an example) and the Vortex System recreates all of the data as it existed at that moment. The basic structure of the Vortex System is shown in Fig. 6.16.

The Vortex System records all disk writes in a chronological sequence, which allows recreation of data as it existed at a past point in time. The system restores file, record, or application data even when the destructive action is more than the infamous (delete *.*). Some examples of this follow.

1. A new member of the accounting department is unfamiliar with the existing package. In order to correct a problem, the accountant initializes the database, which wipes out the wrong records!

2. A new software patch is about to be installed. The network software administrator fails to create a backup of all files before loading the patch. The patch destroys critical records in a custom application.

3. A conversion of a database fails in midstream due to a power failure. Attempts are made to restore from backup tapes for the past two weeks, only to find that the tape subsystem's write head was malfunctioning—all that the LAN manager sees is blank tape.

Since 60 percent or more of all lost data is the direct result of end-user error, plan for it *now!*

To prevent these and other problems the SMS has the capability to back up all of the services shown in Fig. 6.17. Vortex backs up the normal files but also backs up the following:

1. Open files (bind and database)

2. Previous versions in production

3. Name services supported in Netware (DOS, OS2, UNIX, NFS, and MAC)

Banyan Vines is also supported by the Vortex System, in a mirrored environment, as shown in Fig. 6.18. Continuous backup is a future development for the Banyan users and networks.

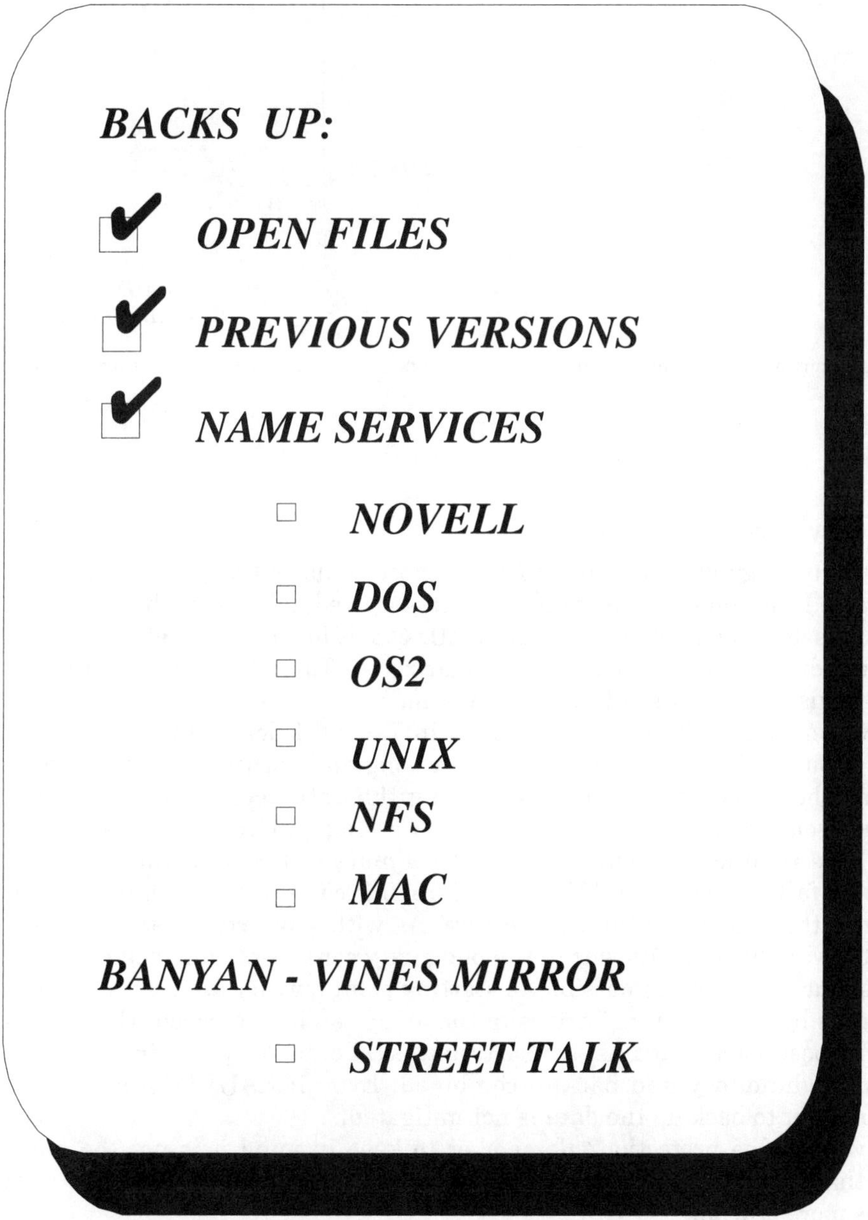

BACKS UP:

☑ **OPEN FILES**

☑ **PREVIOUS VERSIONS**

☑ **NAME SERVICES**

☐ *NOVELL*

☐ *DOS*

☐ *OS2*

☐ *UNIX*

☐ *NFS*

☐ *MAC*

BANYAN - VINES MIRROR

☐ *STREET TALK*

Figure 6.17   Services provided on the Vortex System. (*Source: Vortex Systems, Inc.*)

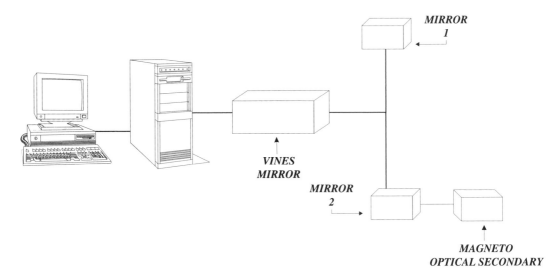

MIRROR
1

VINES
MIRROR

MIRROR
2

MAGNETO
OPTICAL SECONDARY

Figure 6.18 Vines in a mirrored environment adds some protection. Continuous backup is a future development.

RAID (redundant array of inexpensive disks)

RAID technology is also a fault-tolerant process that allows for protection of the data. In an effort to provide the necessary redundancy of the data, this method is drawing a lot of attention. RAID comes in various levels as defined by the industry. These levels are summarized in Table 6.1, which delineates the various components of the RAID system.

However, RAID differs as highlighted in Table 6.1 depending on how it is used. A single drive containing the addressing and parity information could become the weak link. Clearly, spreading the data across multiple drives reduces some of the risks associated with a single drive. When recovering data, this redundant technique preserves more of the data than a single point of failure. Using RAID and mirroring capabilities combined will enhance the chances of data protection. As with any drive, select the best quality available that fits within the budgetary constraints and begin providing as much coverage as possible. A starting point may be two drives; then as time goes by the number of drives in the array can be increased when possible. The best defense against a disastrous loss of data is a proactive approach toward redundancy and backup combined. Even if RAID is provided the requirement to back up the data is not mitigated.

At what value protection? Remember to keep in mind it is not the hardware that is being protected, but the data and lost time, value or productivity that is more crucial.

TABLE 6.1   Levels of Redundancy in RAID Technology

Level	Redundancy description
0	Data is "striped" (spread) across all drives in the array.
1	Mirroring of disks is provided. Critical data will be written to two disks.
2	Distribution of data is done in a bit-by-bit basis. Bits in each byte are spread across the number of disks in the array.
3	One drive provides addressing, parity, and error correction information tables. Data is striped across the remaining drives.
4	Similar to RAID level 3, a single drive stores addressing, parity, and error correction information but blocks are striped across the remaining drives.
5	Blocks of data are striped across all drives in the array. Parity and error correction are distributed across all drives.

Network Recovery

To recover the network after a disaster, the sequence for physical damage would include the following.

1. Cable failure—this has been discussed with the use of temporary wiring or with a wireless option. Find the break point and replace cables from there, keeping in mind distance and signal limitations. A duplicated patch panel can be used on another floor to provide quick reconnection depending on the environment. Some hubs and multistation access units (MAUs) support a dual attached environment that could prove beneficial over time.

2. Electronics—hubs, MAUs, servers, etc., are all simple commodities that can be installed quickly.

3. Physical links—the wires running to the station user or workstation. The cable connection includes the termination of the wires at the workstation end in a standard jack (RJ-45) or connector (BNC, TNC, etc.)

4. Logical links—creating the logical connection to the network implies terminating the cable into a logical port on a server, MAU, etc. Run test and diagnostics to ensure the logical connectivity.

5. Communications equipment—bridges routers, gateways, etc., would now follow the physical and logical connections. Check all connectors, terminators, and screw-down arrangements for continuity, seating, etc.

Communications Recovery

The steps to recover the communications system after a disaster are as follows.

1. Recover the communications server, and power up ASYNC communications servers, modems, etc. FAX servers will also fall into this category. Make the necessary cross-connects to the RJ-11 or patch panels.

2. Communications lines (digital)—if you are using digital lines on the network such as 56 to 64 kbits/s and 1.544 to 2.048 Mbits/s, you should now be

prepared to install them onto the server or other communications device (bridge, router, etc.).

3. Communications (analog) modems that will be used on the network or communications server (pool) should be connected via the RJ-11C connectors.

4. Upon completion of both analog and digital in/out lines establish a connection through each line (both in and out) to ensure that the links are properly working. This would require a full connection across the network from the workstation to a modem across the line, fully connected to a remote device.

5. Once you are satisfied that all of the communication lines, servers, and modems are properly configured and working, the users can begin to access the remote services.

7

Computer Manufacturer's Involvement with LAN Disaster Recovery

Regardless of the plans that a user puts in place, the chance of totally losing access to a building still exists. All of the material discussed to this point has been geared toward a primary site being recoverable. A distinction here is recoverable as opposed to restorable.

- Recoverable implies that in some minimal amount of time, a group of activities can be put into effect at the same location. The normal routine would be that within a 72-hour period, new cables can be pulled or new hardware can be installed in either the existing space or adjacent space in the original site.

- Restorable implies that a great deal of demolition and reconstruction is required to put the original site back into shape. In general, a restorable condition could take months to a year to bring the facility back to an occupancy level suitable for human habitat and network operations.

As a result, the planning process must include a provision of what to do in the event that the facility cannot be used for extended times. This could include the options listed in Table 7.1. Each of the options has merit, yet each also comes with financial implications as well.

Alternative Sites

From Table 7.1 it becomes obvious that the least expensive method to choose is the use of an alternative site. This was discussed briefly in Chap. 5.

TABLE 7.1   Options for Alternative Sites

Option	Advantages	Disadvantages
1. Use existing site.	*a.* This poses limited financial risk. The cost to recover is based on the cost of equipment cabling, etc.	*a.* If the building is not available, then the plan is null and void.
2. Use temporary office in the vicinity.	*a.* Costs are likely to be higher than option 1 above, but space is available. *b.* Temporary space is available in most communities due to overbuilding and recession.	*a.* The space will require prewiring and some security. *b.* In the event of an economic boom, the space may not be available. *c.* If a new tenant rents the temporary space, the whole process will have to be started over. *d.* If a major natural disaster strikes, the alternative building may also be inaccessible.
3. Use a hotel or other transient space.	*a.* Such space is readily available in most parts of the country. *b.* The cost is relatively inexpensive. *c.* Additional space can be obtained as needed.	*a.* Many hotels cater to large convention traffic, and this could limit availability. *b.* The space may have to be prewired by the hotel operator (management). *c.* The space is not totally suited for long-term LAN network operation, but could be made so at a price.
4. Use a hot site vendor either separate from or connected with a computer manufacturer.	*a.* This service is ready to provide data processing services. *b.* The site can be preconfigured to suit individual network needs. *c.* The hot site staff is well versed on getting systems and networks up quickly. *d.* This service buys time to react while operating the network and the day-to-day business.	*a.* This is an expensive option; it usually requires a subscription service. *b.* The stay may be limited to two to three weeks; then the customer must move to a warm or cold site. *c.* If a natural disaster strikes, the hot site may also be affected. *d.* Strong management support is needed for continued funding.

Certain assumptions must be made regarding these alternative sites as outlined in the advantages and disadvantages of each option.

First and foremost is the actual real estate concerns. The LAN manager should not have to deal with this issue. If a real estate department exists

TABLE 7.1   (Continued)   Options for Alternative Sites

Option	Advantages	Disadvantages
5. Use a network hot site vendor.	*a.* The cost is less expensive than a data processing hot site of the cost of a regular mainframe hot site 30 to 35 percent. *b.* The vendor provides LAN wiring and topology to match networks. *c.* This service provides access to network services such as T1, satellite, VSAT (satellite communications using very small aperture terminals), DDS (digital dataphone services, a dial-up data communications arrangement offered by the telephone and long distance companies), etc.	*a.* This service is still more expensive than some options. *b.* The same conditions as in disadvantage 4*b* apply to length of stay. *c.* If more than one disaster strikes, the space could be occupied. *d.* Capacities are usually limited, and the service may not have multiple sites.

within the organization, this can be delegated to them. If this department does not exist, then whoever is involved with renting space will be the appropriate individual. This places the responsibility with an individual or a group that is usually better suited to handle temporary space. The implication here revolves around the ability to

- Get approval from management
- Obtain funding to pay the site costs
- Recruit the real estate department to assist in finding the appropriate space
- Get a commitment from a landlord to reserve the temporary space, and notify the organization if the space will be rented to some other tenant
- Build checks and balances in the plan and the system to alert the LAN manager if any change occurs

Second, the space may have to be pretreated with a variety of LAN safety services. Each of these pretreatments could cost significant amounts of money. Much of the investment may have to be considered "throwaway" costs, especially if the space could be rented to some other tenant. Once the office area is prewired for LANs and treated environmentally, the value of this space increases dramatically. This could be all it takes to get a new tenant to buy into this space. Some of these investments may include the services covered in Fig. 7.1. Either way, the right to protect this investment must be preserved. This is where the real estate department's involvement is needed.

These services, which must be in place in order to move in and set up quickly, could amount to substantial up-front costs. If the choice is to not

1. Prewire station drops (four pairs to the desk)
2. Pull riser cables (100, 200, or 300 pairs)
3. Uninterruptible power supply (servers, workstations, printers, etc.)
4. Fire detection systems (smoke, heat alarms)
5. Fire suppression systems (sprinklers, fire extinguishers)
6. Heating, ventilation, and air conditioning (HVAC)
7. Patch panels in closets (intermediate distribution frames)
8. Telephone connections
9. Miscellaneous

Figure 7.1 Investments needed for alternative sites.

prewire and provide the protection services noted above, the cost goes away until needed. However, the time and effort required to build or provide these services can result in downtime that is longer than anticipated. Using this approach, the disruption could be significant after a disaster strikes. The choices are to take a gamble on the availability of the space or take a longer period to recover if the building is inaccessible.

An example of this is the Meridian Bank fire in Philadelphia in 1991. A fire broke out in a high-rise building in Center City, burning out six floors in the upper portion of the building. By the time the fire was out, a good deal of the building suffered extensive smoke and water damage. After the fire was out the fire inspectors needed to check the building for structural problems. This led to an escalating problem inasmuch as the risk to passersby was a grave concern. Since the heat was so intense the fire officials and the city officials decided to cordon off a five-block square area around the Meridian building. They were concerned that the metal bolts and brackets holding the granite facade on the building might be weakened. If one broke or gave way, tons of granite would come crashing 20+ stories down. The impact would be to send shattered granite shards acting as projectiles in all directions.

Why is this important? Think of how many buildings would be situated in a five-block square area. Each company that occupies space in this area would be scurrying around to find temporary space. Unfortunately, this area was cordoned off for more than three months. Fortunately, Philadelphia had plenty of available office space due to recessionary times; therefore these companies were able to find and occupy space quickly.

How long would it take to set up a LAN on short notice? Given the wiring, environmental treatment, and equipment needs, the delays could be significant, taking as much as a few months. How long can an organization survive without access to the data, particularly critical data?

For these reasons the people responsible for the planning process must consider alternatives. However, the alternative site should be far enough away so that both sites would not be affected by the same disaster. Imagine the impact and embarrassment of having two sites crippled, the primary and the planned alternative site by a single event (such as a hurricane, earthquake, or flood).

What would happen in a city (or town) where space is at a premium? Although it appears that most cities have sufficient capacity to handle growth plans from 5 to 20 years hence, no one can really predict what might happen in the future. A planned alternative site may require a contract to hold the minimal amount of space for the organization. If a new tenant is interested the landlord may be required to give your organization the right of first refusal. The possibilities are many; creativity must be the rule when planning these choices.

Hotels as an Alternative Site

Hotels offer reasonably priced rentals with conference rooms or suites. These could be used as a stopgap in the event building access is denied. By preplanning this with a hotel manager, the odds of getting the "right" type of space will improve dramatically. A conference room (assume a room that could house 30 to 50 people comfortably) could be prewired by the hotel for just these sorts of emergencies. The prewiring could be for both LAN and telephone connections. Usually the hotel will have the capability of performing these functions and has spare capacity in the equipment closets where the prewiring will be run. In the event the hotel has some problems with cosmetic appearances of a conference room being prewired, the list in Fig. 7.2 could be used as a guideline to assist with encouraging a hotel manager. The points addressed in checklist form in this figure are not inclusive but should provide a start.

If hotel aesthetics is still a problem, the hotel could literally hide recessed jacks on the wall 4 to 6 ft high with a flush box cover or a simple picture. See Fig. 7.3 for this alternative, where multiple boxes (using dual RJ-45 jacks) can be hidden behind a simple and inexpensive picture.

Hotels offer a source of easy access for business functionality since they provide access to the services listed in Fig. 7.4. The ability to use each of these services without exorbitant installation fees makes this an attractive

---

1. Pull two 4-pair cables to every drop.
2. Run the cable pair in conduit or flexible metal sheathing.
3. Use level 3 or higher cable (preference is level 5).
4. Terminate the station drops in RJ-45 nonkeyed jacks.
5. Use patch panels in the closet for greatest flexibility.
6. In closets reserve a $3 \times 3 \times 6$-ft space minimum for these connections.
7. If aesthetics are a concern in a conference room, do one of the following:
   a. Coil the cable in the ceiling (least preferable).
   b. Mount the jacks (flush with the surface) on the baseboard.
   c. Install the jacks (recessed) high on the wall at 4 to 6 ft.

Figure 7.2   Checklist for prewiring a conference room.

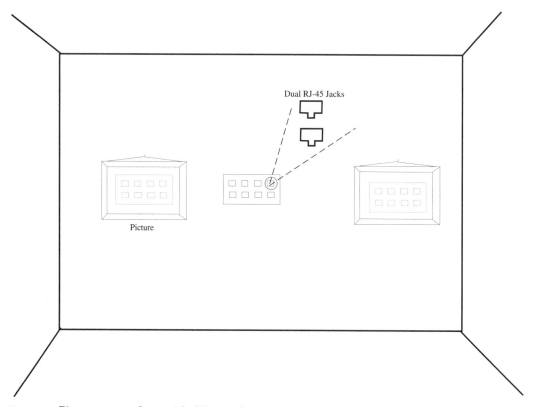

Figure 7.3 Pictures mounted over eight RJ-45 jacks solve aesthetics problems.

alternative site for most organizations. Although some of these items may seem beyond the scope of LAN recovery, one must be aware of all business functions and needs. These are issues that can be provided by the hotel, thereby eliminating the need for the disaster recovery team to coordinate them.

Be cautious when dealing with hotels. Many cater to large convention traffic and may have limited capacities during a large convention. It would be futile to have a hotel as an alternative site and then be denied access due to overcrowding. Furthermore, use one point of contact within the organization to coordinate with the hotel management. Recent events have occurred where hotel managers have agreed to be used as a temporary facility. However, once the various departments (safety, human resources, information services, facilities, and administrative) contacted the hotel staff to review individual needs, the hotel managers became very uneasy and reluctant to proceed. It seems these people were overwhelmed with the responsibility and the possible consequences of these actions. Issues will arise, but one single point of contact can provide calm, manageable demands without frightening hotel managers. There will be some costs associated with this option; however, these should be minimal.

---

1. Telephone
2. Facsimile
3. Sleeping facilities
4. Restaurant services
5. Message taking
6. Voice mail (possible)
7. Pool of labor for moving equipment around
8. Tables and chairs
9. Express services (Federal Express, United Parcel Service, etc.)
10. Parking
11. Air-conditioned space

Figure 7.4   Readily available hotel services.

Computer Manufacturers or Hot Site Vendors as Alternative Sites

Just about every major computer manufacturer has developed some form of support for their customer base. These vendors have developed disaster recovery plans in the past to support both hardware and network failures. Most have a form of recovery site, supporting their own computing platforms and network topologies. All of these vendors charge a fee. They provide a full range of services and capabilities to cover customers' needs during a disaster. Consulting services to help customers develop their disaster recovery plans can also be provided. Note that the consulting assistance will usually point to using a recovery site and subsequent recommendations to use theirs. This is not a disadvantage, but just a point that the user should be aware of. One would hardly expect a vendor to recommend a competitor's services.

To gain acceptance in the LAN recovery marketplace, the computer manufacturers have instituted newer services that accommodate the PC LANs. Some joint ventures have also been used to support a customer with multivendor equipment and network needs. When two or more computer manufacturers team up to provide a single solution, the odds of a smooth recovery at a manufacturer's hot site increase exponentially.

The services these vendors provide include the ones listed in Fig. 7.5. These are generic services; others are available on a site-by-site or customer-by-customer basis. The more complex the need, the higher the subscription fee and the daily fee to use the service during a disaster.

It should be noted that each vendor offers a mix of services that is consistent with the networking architecture that they inherently support. For example, an IBM (International Business Machines) hot site would typically be equipped to support token ring LANs with 327X and 37XX front-end processors, whereas, a DEC (Digital Equipment Corporation) hot site would include the Ethernet-type LAN with a VAX[1] hardware platform and terminal or PC support .

---

[1]VAX™ is a registered trademark of DEC.

1. Floor space, office space (up to 10,000 sq ft)
2. Modems, multiplexers, and data switches
3. LAN wiring systems
4. Topology (token ring or bus)
5. Access to satellite or VSAT service for bridges, routers, etc.
6. T1 links for bridges and routers
7. Front-end processing for access to mainframes
8. Access to dial-up lines (analog at 9.6 kbits/s or digital at 56 or 64 kbits/s)
9. Spare units stocked on site
10. Terminals or PCs
11. Network interface cards
12. Hubs, multistation access units (MAUs), etc.
13. Controllers
14. Consulting assistance
15. Testing assistance

Figure 7.5 Services provided by computer manufacturers.

The fee for these services will be dependent on the amount of space, processing, and capacities required. The complex network will range from $1,000 to $20,000 monthly on a subscription basis depending on individual needs. Normally the site is available to the customer for up to six weeks, at which time a new location is supposed to be provided by the customer, with or without the assistance of the hot site vendor. The hot site is only a temporary stopgap to get the user back into business as quickly as possible while other plans are being put into place.

Network Hot Sites

A new phenomenon in the LAN arena is the use of a network hot site. For years the data processing hot sites have been providing services to their mainframe and midrange customers as already mentioned. This philosophy worked well since all of the emphasis was on large computing platforms. Protection was necessary to provide the comfort level needed by management. The costs were therefore related to the degree of recovery the organization needed.

Now that the LAN has either displaced or been used to supplement many mainframes and midrange computers, the emphasis is focused on protecting the PC network. Therefore the need for large hot sites has shifted to an office environment rather than a data processing shop. These office environments are less expensive than the larger sites and provide different types of services. Many of the services are geared toward the desktop computer plus the use of servers. The amount of LAN disasters on record has spurred this hot site service to allow LAN users to recover to a remote site quickly. This is primarily true when dealing with small to midsize organizations that have limited

resources such as personnel and funding and no additional sites that would typify the larger organizations. These network hot sites can be the small organization's salvation after a disaster. (A small company can be typified by having a
single LAN with up to 50 nodes, at a single site. Midsize companies have a single or multiple LANs with up to 200 nodes, at multiple sites. Large companies
use multiple LANs with greater than 200 nodes, at multiple sites.)

Picture the impact on an organization after an event such as hurricane
Andrew in Florida. Think of all the small and midsize companies that have
been crippled due to the loss of their buildings, loss of LAN equipment, and
the inaccessibility of their data. How many will survive this form of disaster?
Conversely, how many will cease to exist as a business entity? Here a network hot site, located in another city, could be the appropriate recovery
method. By moving in on a short-term basis, these organizations could
reestablish their functionality and serve their customers. Admittedly, if the
business serves only local Miami customers, the use of a hot site may have no
benefit, or, if it is a manufacturing plant that could not be replaced to produce products, the hot site may add little to the recovery.

However, as the industry has evolved, the need to handle customers from
service businesses has magnified. Therefore the access to data and PCs to
manipulate the data is far more essential. Consequently, these network hot
sites become more valuable to the organization.

To understand what these organizations provide, Fig. 7.6 lists the services for
data communications support. This support would be crucial if the hot site is
located in another city, state, etc. Access to the LAN-based services from remote
offices would be dependent on the data communications aspect of the network.

Note that much of these capabilities exist also at the major hot site vendor's locations. Other capabilities these network hot sites may typically provide include network management services to support various LAN operating systems. These are included in Fig. 7.7. With the network management
systems, these hot sites also attempt to support the typical network management software systems, which are also included in Fig. 7.7.

Understandably, these are the more readily available operating systems
and software support services. In the event a particular client has a different
system, the hot site vendor would likely purchase and/or support the system.
Most of the different versions would be handled on a case-by-case basis. Here

---

1. Modems at 2.4 to 9.6 kbits/s and analog dial-up lines
2. Data switches and matrix switches for connectivity
3. Front-end processors (37X5)
4. Switched 56/64 kbits/s access
5. T1 lines
6. Reserve T1 lines
7. Switched T1 access (optional)

Figure 7.6 Data communications services at network hot sites.

1. Systems
   a. LAN Manager
   b. Novell 2X, 3X
   c. UNIX
   d. BANYAN
2. Management software
   a. Netview
   b. LAN manager or server
   c. Novell
   d. OSI (open systems interconnect)
   e. SNMP (simple network management protocol)

Figure 7.7 Network management services.

is where the hot site vendor will provide the service on a cost plus service fee structure that could be far more expensive than the more generic or standard systems. The organization must decide what level of support to provide for themselves and how much the hot site vendor will be expected to provide. This becomes a business decision based on talent availability and finances.

The larger network hot site vendors have become more proactive in assisting their customers in planning for a disaster. They cover the non-LAN issues that a typical LAN manager may never think of. For example, a LAN manager's primary emphasis is to recover the hardware, operating software, and the actual data. But when relocating to a hot site that is remote from the organization, human logistics must be considered. Larger companies that use remote sites involve several departments in the planning process. These might include

- Human Resources
- Safety and security
- Facilities
- Travel
- Finance
- Purchasing
- Information services
- Telecommunications

Each of the above departments would be responsible to carry out certain duties in support of the recovery process. The goal is to get back into business as quickly as possible.

Consequently these issues must be addressed, regardless of who shoulders the responsibility. The small to midsize organization may have none of these support staffs to draw upon. To overcome this problem the hot site vendor could provide the human logistics support as shown in Fig. 7.8.

1. Family considerations—providing child care facilities for workers who have small children and care of sick or invalid members of the household.

2. Transportation arrangements—either air transportation to move the staff to the hot site at any hour, or ground transport at the remote sites, such as van or car rental, etc.

3. Shelter and food accommodations—arranging for hotel lodging and setting up accounts to bill room and food charges.

4. Bank transfers—setting up electronic fund transfers to a local bank for daily operations; arranging for limited withdrawals for employee sending cash.

5. Coordination of message center services—allowing message notification for displaced workers.

6. Medical and health care—arrangements for physicians and prescriptive services as needed.

7. Documentation and clerical services—begin the documentation of actions taken, costs, expenditures, etc; provide temporary clerical help to input data as necessary.

8. Other services as needed.

Figure 7.8   Areas covered by the hot site vendor.

Although these issues are not high on a priority list for a LAN manager, they certainly are important. After a disaster the users' primary role is to get the LAN and the organization back in business. However, if their personal needs are not being handled, they cannot concentrate on doing their normal jobs. They need to be assured that all is well before they can focus all their energies into the recovery.

Chapter

8

Writing
the Plan

Getting this far in the process, the following materials have been addressed:

1. Understanding the importance of LAN disaster recovery planning
2. Conducting the business impact analysis
3. Presenting the business case to management
4. Considering the alternatives to preparing the plan
5. Protecting the LAN from various day-to-day risks
6. Protecting the communications portion of the LAN
7. Recovery procedures: both physical and logical
8. Backup and redundancy options
9. Alternative site capabilities

The Next Steps

What is the next step? Actually, throughout this book, we have been addressing the procedural issues of the action plan that must be provided. In each of the previous chapters a particular portion of the plan has been covered. The next step then is to put it all on paper. Using the checklists, tables, and written material provided, the plan can now be finalized. No one plan or format will fit every organization. However, the form and format may already exist in an organization. If a disaster recovery plan exists within the organization, whether it is for data processing, facilities, or another department, use this as the primary format to provide consistency of the documents. A better way to use an already existing plan is to create connections or references to it from the LAN disaster recovery plan.

In Fig. 8.1, a master document exists: the company disaster recovery plan. From this master document a data processing plan, a human resources plan, and a LAN plan are all developed. Each plan has references built into the

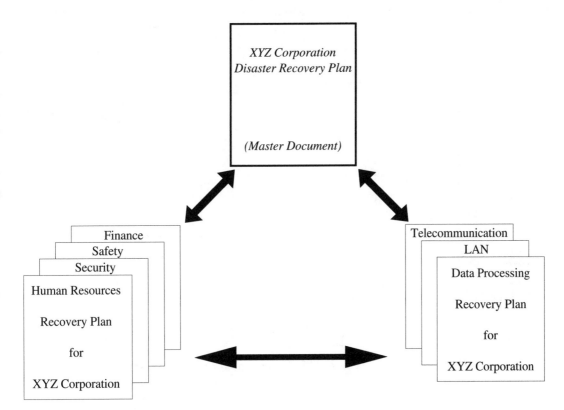

Figure 8.1 Each of the subplans must tie to a master plan.

other plans, with all subplans supporting the master plan. It would be ill-conceived to have a recovery plan strictly for the LAN environment if an organized set of activities to recover the rest of the business did not exist. For example, a LAN can be brought up at a recovery site (or on the same site). If no one is organizing the activities of the personnel pool, then chaos will still reign and nothing would get done. Furthermore, if the LAN equipment is running but the physical environment (such as power, HVAC, lights) is not addressed, the space will be useless and unsuitable for human occupation. These activities must be considered together or the total function of running a business will be lost.

As the plans are being written from the master document to each of the subplans, a direction statement must also be provided. This direction statement comes from the upper echelons of management in the form of the one-, three-, and five-year strategic business plans. It would be futile to attempt to write a plan (which could take six months to two years) that will become a living, breathing document if the plan developers have no idea about the organization's goals and future directions. This document (the disaster recovery plan) could be going in one direction, whereas the business is headed in a different direction, as shown in Fig. 8.2. When called upon to recover a LAN

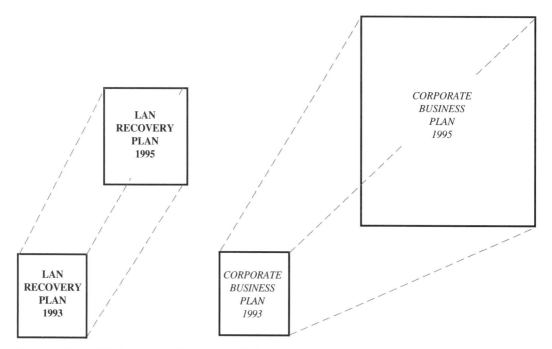

Figure 8.2   The LAN plan may not keep pace with the business growth plans, thus making it useless.

separately or as part of a more global recovery, the plan could be obsolete and thus useless.

This is a difficult piece of the planning process. The LAN manager is typically focused on day-to-day problem solving, which may involve installation of new equipment, resolving hardware problems, solving software glitches, or fighting for daily funding. Taking a longer-range view of the organization's needs may be simply an impossible task. Getting the time to forecast rather than react to user demands may be a monumental endeavor, so the issue becomes more clouded.

Writing the plan (seriously looking at all of the possible risks coupled with future growth, standards, and budgetary needs) is a long-term and time-consuming process. There are very few LAN managers that can afford the luxury of taking four hours per day to devote toward developing such a plan. Consequently, as day-to-day failures and panic situations arise, the plan gets moved to the back of the manager's mind or the bottom of the priority list. However, when we look at all the tasks to be addressed by a plan, one person's efforts make this a very lengthy process. In most cases the timing is longer than I've projected just by the tedium involved and the interdepartmental dependencies. This is considered a "necessary evil" rather than a planned perpetuity approach. Pity the LAN manager who is supposed to be the technical guru for hardware, software, cable, and security issues, now being thrust into a strategic planning role. In Table 8.1 a

TABLE 8.1   General Timetable to Develop a Plan For a Midsize Organization

Number	Tasks	Time
1	Develop a list of risks to the LAN	4 weeks
2	Conduct a business impact analysis	3 to 6 weeks
3	Present case to management	2 weeks
4	Recruit assistance from within the organization vendors and carriers	4 weeks
5	Inventory all equipment and cabling systems	8 to 12 weeks
6	Verify software licensing and the number of users on the LAN	2 to 4 weeks
7	Label all cables, connectors, and hubs	6 to 8 weeks
8	Secure physical rooms, closets, etc.	1 week
9	Develop a standards document for all present and future systems: purchases of equipment, backup systems, etc.	4 to 6 weeks
10	Develop action plans	10 to 12 weeks
11	Document everything	16 weeks
12	Order new hardware, software, and miscellaneous components	1 week
13	Install new systems	3 weeks
14	Coordinate off-site recovery locations	1 week
15	Coordinate off-site storage	1 week
16	Develop and train all staff, management, and recovery teams	8 weeks
17	Develop a test plan and establish necessary frequency schedules	6 weeks
18	Change control and documentation	Ongoing
	Total time	80 to 95 weeks

general timetable is presented showing some of the appropriate tasks and associated times required for a midsize organization. The midsize company has up to 200 nodes (PCs). This is not a very large organization. Therefore the luxury of "extra" staff is nonexistent. Likely there will not be a specialist in facilities, electric, cabling, HVAC, etc., as you would find in the Fortune 500 companies. This puts more pressure on the disaster recovery planner. If the LAN manager is a staff of one, as is usually the case, this will be next to impossible to accomplish. Yet, it must be done! If not, the business survivability is placed at risk, particularly with critical operations running on such an exposed network.

Although 80 to 95 weeks may seem inordinately long to get to a plan that is useable, one must consider the various coordination efforts with other departments, the constant changes that will occur requiring rewrites and updates, and the general lack of solid blocks of time to conduct these activities. Those who can accomplish these tasks in six to eight months are ahead of schedule and deserve applause. Reality, however, says that the above-noted times will be more appropriate. However, caution must be placed in assigning a start date. Managers who start a project that will take 80 to 95 weeks require a lot of discipline to stay on target and within budget. As part of human nature, managers adopt a procrastination philosophy that they have 80 weeks to get the job done, so they can put it off for a while. Unfortunately time slips by quickly and the project never gets done. Worse yet, as the time grows short, a haphazard plan gets slapped together just to meet the time frames. This self-imposed pressure yields a poor plan that potentially would never work and jeopardizes the LAN manager's credibility in management's eyes. Either case is devastating to one's career potential. Take the time to do it right! Do not get caught up in the daily routine, which will leave no time for strategic planning. This is a critical need for the business.

LAN Standards

The subject of standards is one that receives very little attention in many organizations. Yet it is definitely an area that can help prevent problems in the future. Management must be supportive and produce a statement to the organization. This statement will deal with the following critical issues:

- Equipment purchases, policy, and procedures
- Software systems
- Policy statements regarding software piracy
- Security (physical and logical)
- Virus prevention
- Access control and enforcement
- Telecommunications access
- Backup systems and the required frequencies of doing the backup
- Operational support systems

A sample of a LAN standards table of contents is contained in Fig. 8.3. This is not an all-inclusive table of contents but provides a shell for each organization to customize as they see fit. Remember, no one document meets every organization's needs, so this becomes an internalized vehicle. Use what best fits the needs and change what does not fit. The best plan is one that is written within the organization given an understanding of how the internal operation works.

Purchasing and acquisition

These standards cannot be overemphasized; they provide a framework for both the daily activities and the provisions required in the event of a disaster. In these standards documents, responsibilities will be assigned that regulate purchasing of new systems, components, and applications. The standards are not designed to inhibit users from acquiring hardware and software to meet a business need. Instead they offer the ability to purchase solutions that will migrate onto the existing networks and systems without creating isolated pockets of applications. Where hardware conflicts might

Table of Contents

I.   Objective: Defining mission critical needs
        Mission critical needs
        Non-mission-critical needs
II. Security
        Physical
        Environmental
        Logical
III. Operational support issues
        Purchasing policies and procedures
        Documentation
        Backup policies
IV. Access control
        Passwords policy
        Change control policy
V.   Virus protection
VI. Software licensing, piracy, and enforcement
VII. Telecommunications access
        Remote access
        Dial in or dial out

Figure 8.3   A sample table of contents for LAN standards.

arise, the purchase can be postponed until a suitable product can be found. This prevents a system from being introduced onto a LAN that will cause problems.

Part of the standards document may well support a research and development (R&D) LAN (a test bed) where all new hardware and software can be installed, configured, and tested prior to being introduced on a production LAN. Again this is a protection methodology rather than a restrictive philosophy.

By forcing the purchase of hardware and software through a focal point, perhaps the LAN manager, licensing information can be logged, documented, and tracked. This will help with the prevention of piracy. Periodic inventories of all applications on workstations and servers should be conducted, where the licensing can be reverified. Any software found that has been installed without the proper documentation should be instantly flagged and deleted as appropriate. Enforcement should include a report to management of all violations to the policy.

Inventories

To conduct inventories several software packages exist on the market that expedite the process. These applications go out across the network to verify all hardware components such as the following:

- PC type (286, 386, 486)
- Printer attached (parallel or serial)
- Graphics cards
- Disks (floppy and hard drives)
- Network cards installed with appropriate address
- Additional peripheral devices (modems, tape systems)

A database of each component is then built from this information. Cross-references of serial numbers, devices, and configurations can be built to show the connection point (jack) to the closet (patch panel or hub) through the network.

As the hardware is checked, many of the utility inventory programs also scan the drives on each PC, workstation, or server for the software and applications that are installed. Once again this can be used to build a database for future reference. Licenses can be verified, documentation can be maintained, and any changes can be flagged for follow-up.

Virus prevention

To prevent the risk of viruses, all new software should be checked first with a detection and correction program. Myriads of these applications exist, but a single product that provides consistent checks and regular updates may be more appropriate. One cannot assume that software purchased from any vendor is virus-free. No matter how reputable or how large the company is, risks of infected applications exist everywhere. A standard requiring that checks are run on all software by the LAN staff prior to being loaded is obviously the best philosophy.

To overcome the risk of infection, many organizations have removed floppy disk drives from the workstation. A bootable prom (programmable read-only memory) is used to attach the device to the network rather than allowing the user to boot from floppy disks, etc. Although this may seem drastic, it is a means of preventing a very real threat from viruses being introduced on the LAN and disrupting many other users.

Passwords

Passwords are always a sensitive subject in any user environment. Management must support and document in the standard that passwords will be used. Furthermore, the frequency of password changes must be stated, whether it is monthly, quarterly, or whatever cycle works best.

As these standards are distributed, users get a very clear message: Management supports the LAN manager's role and violators will be dealt with. This gets the planning for recovery well on the way by being proactive in defining how the system should work. The result is a prevention program that will eliminate some risks and mitigate others. Use this LAN standards document wisely to protect the organization's assets. It can go a long way.

The Plan

The plan is the next phase after the standards document is developed. This is the working document that culminates with the action items and sequences to be followed in the event of a disaster. One of the simplest forms of a plan is the informal plan that resides in the LAN manager's head. This is also the most dangerous. What if the LAN manager leaves the organization? What if he or she is not available and a disaster strikes? Who else will know what to do and when and how to do it?

Thus, the *written plan* is a true necessity. When developing the plan, remember to keep the master document in mind and work from there. Use the indexed references to the master document whenever necessary. Write action plans on how to recover the LAN. Keep it short and simple (KISS).

Do not reinvent the cycle and the documentation process. Use the tools that are already available within the organization.

A sample table of contents for the LAN disaster recovery plan is shown in Fig. 8.4. This is a general outline that can be followed or used as a shell that can be modified. In this table of contents the plan is broken down into five major categories, as shown in Table 8.2. These five categories provide the infrastructure of the plan where the individual components are then added as needed.

Forms

As the plan is being developed, a series of forms are used. Many of them have already been highlighted throughout this book. However, others may prove useful, either from the master document or the LAN disaster recovery plan. Some additional forms are included here, which can be used as samples. In any case these can be modified as needed to suit the needs of the individual organization.

In Fig. 8.5 a sample checklist that covers specific backup requirements is shown. This checklist can be used to formulate the backup policy and verify its continued operation. Figure 8.6 is a checklist for hardware recovery with minimum service levels provided at a remote site (hot site, alternative site, network hot site). This form deals with providing equipment and communications links necessary to get back into business quickly. Figure 8.7 is a form establishing a list of personnel needed for team formulation to provide the appropriate staffing and key on training needs for the staff. Figure 8.8 is a sample of a phone tree for notification of teams, etc. In this form a primary team member and one or two alternates are selected. This is optional, as the size of the organization will dictate the availability of teams and members. One point that should always be considered, however, is that no one person should be listed in more than one place. It is crucial that responsibilities be assigned to get things moving quickly. One person cannot be in multiple locations at the same time.

Figure 8.9 is a sample bomb threat checklist. Many people are trained in receiving bomb threats and understanding what must be learned from the

Table of Contents
LAN Disaster Recovery Plan

I. Administrative section
   A. Purpose of plan
   B. Scope
   C. Management letter of commitment
   D. Goals and objectives
      1. Protect human life
      2. Minimize risk to organization
      3. Minimize downtime
      4. Maximize return to business
      5. Prepare to recover critical operations
      6. Protect against legal action
      7. Preserve customer and employee confidence
      8. Assess business impact
      9. Identify mission critical applications
      10. Define recovery and restoration strategies
II. Action plans
   A. Policy statements
   B. Timing of recovery efforts
      1. Within 2 hours
      2. Within 72 hours
      3. Within months
   C. Procedures to follow
      1. Loss of building
      2. Power interruptions
      3. Loss of life
      4. Servers destroyed
      5. Bridge, router, or gateway failure
      6. Communications failure
      7. Threats of hackers
      8. Sabotage
      9. Bomb threat
      10. Natural disaster (fire, flood, tornado, etc.)
      11. Physical security breaches
      12. Cable cuts
      13. Major data loss
      14. Hardware and software crashes and glitches
   D. Initial actions
      1. Prevent damage
      2. Recognize problem
      3. Assess damage
      4. Notify management
      5. Notify disaster recovery team
      6. Notify vendor and carrier
   E. Follow-up actions

Figure 8.4 Table of contents for a LAN disaster recovery plan that can be used as a sample.

Table of Contents
LAN Disaster Recovery Plan (*Continued*)

1. Mitigate damage
2. Begin initial recovery
3. Notify alternative site
4. Notify data storage site
5. Notify employees and management
6. Coordinate with local authorities
7. Maintain security
8. Notify insurer
9. Begin documentation of event

F. Business resumption
1. Activate full recovery team
2. Reroute network
3. Make the decision to declare disaster
4. Relocate to alternate or hot site
5. Bring up hardware
6. Bring up operating systems
7. Attach to network
8. Recover data from backup

G. Rerouting of network facilities
1. Recover control desk
2. Activate help desk
3. Relocate and/or replace bridges and routers
4. Connect lease lines
5. Recover dial-up services
6. Verify operations

H. Concurrent activities
1. Log all events
2. Order new equipment and software
3. Increase physical security at original site
4. Enforce physical security at recovery site
5. Change passwords

I. Restoring critical functions
1. Demolition activities at old site
2. Uninterruptible power supply (UPS) and power
3. Heating, ventilation, and air conditioning (HVAC)
4. Fire detection and suppression
5. Cabling systems
6. New hardware installed
7. Operating systems configured
8. Applications loaded
9. Operations tested and verified
10. Migration of workstations and servers begun
11. Users migrated back onto LAN

Figure 8.4 (*Continued*)

Table of Contents
LAN Disaster Recovery Plan (*Continued*)

J. Documenting actions
    1. Review all activities logs
    2. Assess vendor and carrier performance
    3. Prepare after action report
    4. Assess if new procedures needed
    5. Keep management informed
K. Recognition
    1. Recognize achievements
    2. Schedule time to recoup
    3. Plan awards and recognition party
III. Testing and Maintenance Strategies
    A. Testing policy statement
    B. Maintenance policy statement
    C. Responsibilities assigned
    D. Cross-references to other documents
    E. Procedures
        1. Frequency of tests
        2. Coordination
        3. Types of tests (scheduled and nonscheduled)
        4. Introduced complexities
        5. Results
        6. Changes required
    F. Maintenance of documents
        1. Frequency of updating
        2. Choice of responsible person to maintain documents
        3. Change control management
            a. Hardware
            b. Software
            c. Team members
            d. Communications methods
            e. Vendors
        4. Delivery method
        5. Periodic reviews
        6. Audits
IV. Training
    A. Training policy
    B. Scope and degree required
    C. Responsibilities assigned
    D. Cross-references to other documents
    E. Frequency
        1. New team member
        2. Management
        3. Recovery team

Figure 8.4 (*Continued*)

```
┌─────────────────────────────────────────────────────────────┐
│                    Table of Contents                          │
│            LAN Disaster Recovery Plan (Continued)             │
│         4.  Specialty team                                    │
│         5.  All employees                                     │
│         6.  Remedial                                          │
│         7.  New equipment and applications                    │
│   V.  Appendixes                                              │
│         A.  Phone tree for management and recovery teams       │
│         B.  Recovery team duties                              │
│         C.  Inventory and report forms                        │
│         D.  Forms detailing control of changes               │
│         E.  Inventory of hardware                             │
│         F.  Software licenses                                 │
│         G.  Vendor and carrier phone tree and escalation lists│
│         H.  Diagrams and schematics of LAN                   │
│         I.  Diagram and schematics of communications lines    │
│         J.  Others as needed                                  │
└─────────────────────────────────────────────────────────────┘
```

Figure 8.4 (*Continued*)

TABLE 8.2   Five Major Categories for a Plan

Category	Description
I	Administrative section—provides overall theme and policy statements
II	Action plan—provides the details of what should be done
III	Testing and maintenance strategies—details the procedural policies
IV	Training—provides the levels and frequency of training needs
V	Appendixes—details of various other components as needed

caller. However, when a call actually comes in, it is unexpected, leaving the recipient in a panic situation. This checklist helps the user fill in the blanks so that logically gathered information can be obtained.

Figure 8.10 is a telephone notification form to verify the disaster call. After receiving a call off-hours, unexpectedly, the recipient should jot down specific information. Then the recipient should place a call back to the appropriate location and verify the information. This rules out prank calls and maintains control of the situation.

Figure 8.11 is a flowchart that can be used to organize the disaster activation and arrangements for the meeting place. This is a decision tree that flows with the actual notification process. Figure 8.12 is a personnel change notification to provide for additions or deletions from teams. In this case, a deletion cannot be made unless a subsequent addition is also submitted. However, additions can be made at any time to expand a list.

Figure 8.13 is an organization chart, taken from a master document, that shows the recovery team coordinators. In this particular form, the notification process begins with the appropriate department coordinators, who in

BACKUP PROCEDURES CHECKLIST

1. Is a central library being maintained for all software licenses and program disks?

    ☐ yes   ☐ no

2. Are backups being created routinely?

    ☐ yes   ☐ no

3. What frequency are backups run?

    ☐ daily   ☐ weekly

4. How often are tapes rotated off-site?   ☐ daily   ☐ weekly   ☐ monthly

5. Does an archiving program exist?

    ☐ yes   ☐ no

6. Does it produce a catalog of tapes/files?

    ☐ yes   ☐ no

7. Is security maintained by file? Directory?

    ☐ yes   ☐ no

8. Is change control procedure used for all software modifications?

    ☐ yes   ☐ no

9. Do procedures exist for handling of tapes/disks? Is a person designated to control?

    ☐ yes   ☐ no

10. Are separate people responsible for creating backups and off-site handling?

    ☐ yes   ☐ no

11. Are confidential backups handled differently?

    ☐ yes   ☐ no

12. Are all mission critical files and applications labeled and listed?

    ☐ yes   ☐ no

Figure 8.5   A sample checklist establishing the tape backup procedures.

NETWORK RECOVERY HARDWARE CHECKLIST
FOR HOT-SITE LOCATION

1. Hardware configuration at original site and recovery site
   # PCs _____ PC Family (AT, 386, 486, PS/2 etc.) _____

2. Tape Backup system _____
   (Manufacturer and capacity)

3. Network Topology _____
   (Ring, Star, Bus)

4. Network Cards _____   _____
   (Make, Model #)   Qty.

5. Cable System _____   Jack _____
   (Twisted Pair, Coax, Fiber)   (RJ-45, Balun, BNC)

6. Closet Equipment _____   _____
   (Concentrator/Hub etc.)   (Make)   (Model)   Qty.

7. Special Equipment _____   _____
   (Scanners, Barcode, Modems)   (Make)   (Model)   Qty.

8. Printer Needs _____ _____ Interface _____
   (Make/Model)   (Qty.)   (Serial, Parallel)

9. UPS System _____ _____   _____
   (Make)   (Model)   (Size, Capacity)

10. Documentation on Hand   ☐ Yes   ☐ No

Figure 8.6 Minimum hardware requirements at a hot site.

turn notify their respective teams. The human chain should never be broken: if a primary team member is not available, the phone call should proceed to the alternate. Note that the LAN team is highlighted with specific team titles.

Figure 8.14 is an inventory and critical rating form. During the inventory process, the teams interview department managers to determine the criti-

TEAM FORMULATION CHECKLIST

RECOGNITION OF TEAMS:	NUMBER OF PERSONNEL NEEDED	SPECIAL SKILLS REQUIRED
1. Fire suppression teams	_____	_____ (Fire equipment handling)
2. Human resources team	_____	_____
3. Safety/Security team	_____	_____
4. Vendor personnel teams	_____	_____
5. Hardware teams	_____	_____
6. Software team	_____	_____
7. Hot-Site team	_____	_____
8. Help desk	_____	_____
9. Telecommunications team	_____	_____
10. Facilities team (electrical/HVAC etc.)	_____	_____
11. Medical/First aid team	_____	_____ (First Aid, CPR)
12. Cabling team	_____	_____
13. General clerical	_____	_____

Figure 8.7 Team formulation for disaster recovery includes the number of personnel and any special skills required.

cality of each device. A cost and serial number is associated with each device. Ratings would be on a 1 to 5 basis, with 1 being most critical and 5 being least critical.

As mentioned in Chap. 2, a business impact analysis is used to justify the disaster recovery plan to management. The work sheet shown in Fig. 8.15 compares the options of damages when a plan is not in place to the damages that can be expected when a plan is in place. This becomes an effective presentation tool when presenting a business case to management.

Furthermore, when looking at the risks involved with a LAN, some probabilities of major events are easier to plot than others. Figure 8.16 is a checklist used to develop a low- to high-risk evaluation on the systems, based on particular events. The probabilities or events can be modified to suit the individual organization's needs.

Other forms can be created, depending on the specific needs assessment and applications that are installed on the LAN. The forms above are designed to get the process underway and in the right direction. Note that many of these forms and checklists are geared to collect only a small amount of information that can be easily turned into a graphic representation for

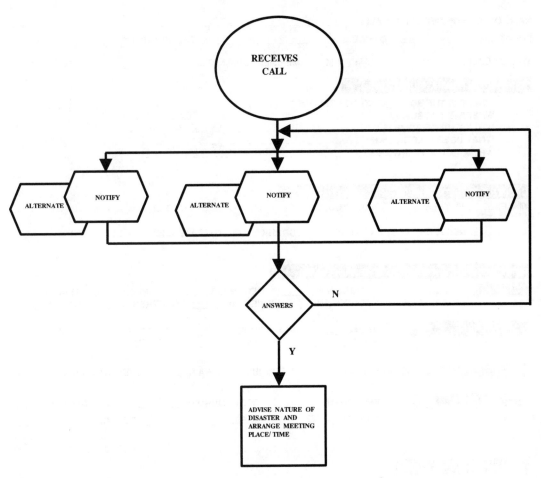

OFFICE CALLING CHART

Figure 8.8 A phone tree is used to contact team members, either primary or alternate.

management. A narrative is not needed if a picture can say the same thing. This format should be used throughout the entire planning process. Use graphics or checklists that are easily visualized and filled in. The use of graphics as opposed to narrated text will expedite a recovery process. Lengthy documents that are unnecessarily verbose could impede the recovery process. A user or recovery team should be able to pick up the action plan and recover the network without reading a lot of preparatory text. By using the simple flowchart or graphic, the process will run smoother, and the chances of recovering will quickly increase exponentially. Remember, keep it short and simple (KISS).

BOMB THREAT CHECKLIST

NAME OF PERSON RECEIVING CALL: _____

LOCATION: _____ OFFICE: _____ TELEPHONE NO: _____

TIME OF CALL: _____ ☐AM ☐PM   TIME CALL TERMINATED: _____ ☐AM ☐PM

ASK THE FOLLOWING QUESTIONS:

    WHERE IS THE BOMB GOING TO EXPLODE? _____

    WHERE IS THE BOMB? _____

    WHAT KIND OF BOMB IS IT? _____

    WHAT DOES IT LOOK LIKE? _____

    WHY ARE YOU DOING THIS? _____

CALLER DESCRIPTION: (CHECK ALL THAT APPLY)

APPROX.

MALE: _____ FEMALE: _____ VERY YOUNG: _____ YOUNG: _____ OLD: _____ AGE: _____

CALM: _____ NERVOUS: _____ EXCITED: _____ DRUNK: _____ THREATENING: _____ OTHER: _____

VOICE DESCRIPTION: (CHECK AS APPROPRIATE)

ACCENT: NONE:_____   FOREIGN:_____   REGIONAL:_____   SPANISH:_____   BLACK:_____

    EASTERN:_____   EUROPEAN:_____   ORIENTAL:_____   SOUTHERN:_____   OTHER:_____

CHARACTERISTICS: NASAL: _____   HIGH PITCHED:_____   LOW PITCHED:_____   LISP:_____

    SLUR: _____   RESONANT: _____   DEFECT: _____

CALL ORIGIN: SOUNDS NEAR: _____   SOUNDS FAR AWAY: _____   OPERATOR ASSIST: _____

BACKGROUND NOISES: PHONE BOOTH: _____   STREET SOUNDS: _____   BABY CRYING: _____

    LAUGHTER: _____   MACHINERY: _____   VOICES: _____   OTHER: _____

EXACT WORDS OF CALLER:

NOTES:

Figure 8.9   The bomb threat form is used to gather information about the caller.

TELEPHONE DISASTER REPORT FORM

DATE: _____   TIME: _____

NAME OF PERSON CALLING: _____

TELEPHONE NO. OF CALLER: _____

ORGANIZATION OF CALLER: _____

DESCRIBE THE NATURE OF THE EVENT: _____

_____

_____

_____

_____

_____

ACTIONS TAKEN:

☐ RETURNED CALL TO VERIFY INFORMATION

☐ DETERMINED IF ERT ACTIVATION NECESSARY

☐ CALLED BUILDING SECURITY ( )_____
  PERSON SPOKEN TO: _____

☐ BUILDING ACCESSIBLE ?

  ☐ YES   ☐ NO

☐ NOTIFIED DISASTER RECOVERY
  COORDINATOR AT: _____

☐ NOTIFIED MANAGERS OF
  _____
  _____
  _____

☐ NOTIFIED OTHERS: (CARRIERS, VENDORS,ETC.)
  _____
  _____

☐ ARRANGED MEETING PLACE AND TIME
  PLACE: _____
  TIME: _____

DESCRIBE ANY OTHER ACTIONS TAKEN:

_____

_____

_____

_____

Figure 8.10   The telephone notification form is used to gather and verify disaster call information to preclude a hoax.

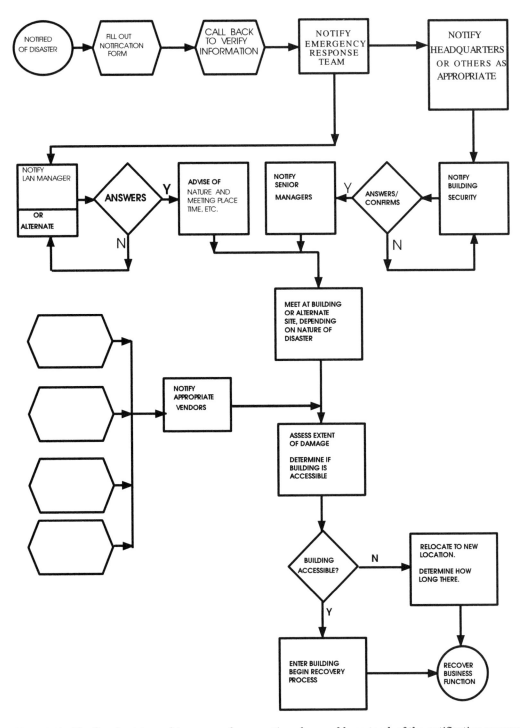

Figure 8.11 The flowchart is used to arrange for a meeting place and keep track of the notification process.

CONFIDENTIAL

PERSONNEL CHANGE NOTIFICATION

DELETIONS

TO: _____

FROM: _____

The following individuals' names should be deleted from the Disaster Recovery notification checksheets from the positions noted below. These changes should be made immediately.

	Name:	Department:	Position on Team:	LIST:	Telephone numbers			Reason for Change:
					Home	Office	Pager	
1.								
2.								
3.								
4.								
5.								

SIGNATURE: _____ DATE: _____

- *FOLD HERE* - -

Note: For each deletion on a list above, an addition must be created below

---

PERSONNEL CHANGE NOTIFICATION

ADDITIONS

TO: _____

FROM: _____

The following individuals' names should be added to the Disaster Recovery notification checksheets in the positions noted below. These changes should be made by: _____

	Name:	Department:	Position on Team:	LIST:	Telephone Numbers			Replaces:
					Home	Office	Pager	
1.								
2.								
3.								
4.								
5.								

SIGNATURE: _____ DATE: _____

CONFIDENTIAL

Figure 8.12 The personnel change form for team members. This form is used for additions and deletions to the team structure.

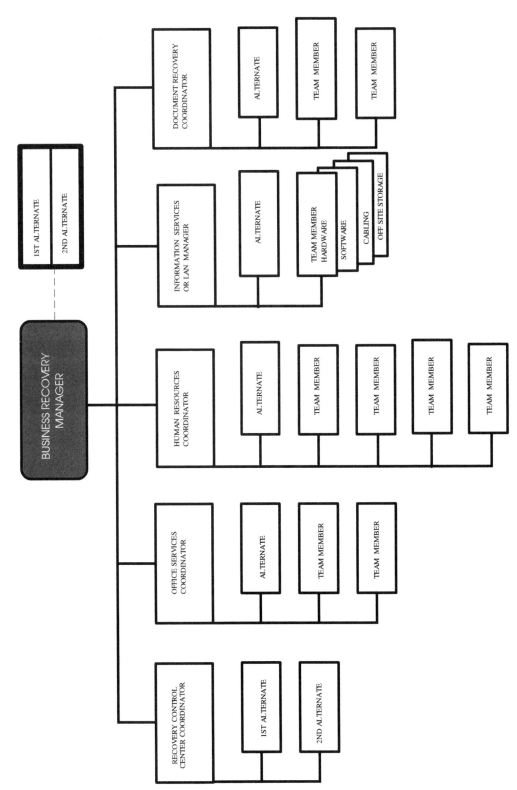

Figure 8.13 The organization of a team is shown in easy to use formats. The primary team member and at least one alternate are listed.

INVENTORY AND CRITICAL
RATING OF EQUIPMENT FORM

ITEM:	COST:	SERIAL #:	LOCATION:	USER:	RATING:

Figure 8.14 The inventory and a critical rating are done on all applications and hardware systems on the LAN.

JUSTIFYING DISASTER RECOVERY PLANNING
BUSINESS IMPACT ANALYSIS:

	NO ACTION PLAN			WITH ACTION PLAN					
EVENT:	PROBABILITY	LENGTH OF OUTAGE	COST OF LOSS	PROBABILITY	LENGTH OF OUTAGE	COST TO RECOVER	ONE TIME COST	PRIORITY	TIME TO IMPLEMENT

Figure 8.15   Comparing costs with or without a plan and impacts.

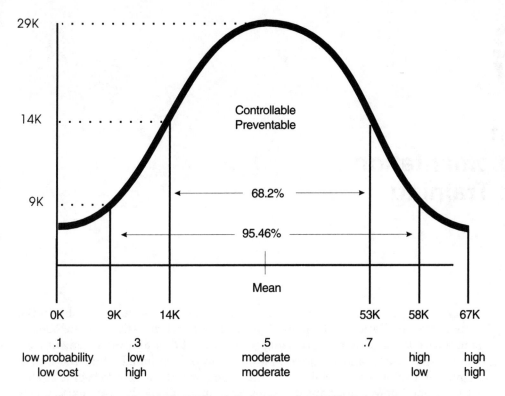

0K	9K	14K		53K	58K	67K
.1	.3		.5	.7		
low probability	low		moderate		high	high
low cost	high		moderate		low	high

A business impact analysis can be plotted onto a standard bell shaped curve to represent the information from the table.

Quadrants are divided into:

Probability:	Low	Medium	High
Cost:	Low	Medium	High

Medium

High		High
	Low	Low

Figure 8.16   Checklist to develop a risk evaluation.

9

Plan
Implementation
and Training

Congratulations! Now that this point is reached, the next phase is to begin the actual implementation of the plan. Several charts and checklists are included to summarize these activities. By using some form of a project management software, the tracking and resource planning can be simplified. Once the implementation begins it is crucial to stay on target, or if the schedule begins to slip, let everyone know. Important decisions were made along the way, particularly at upper levels of management. These decisions include the following.

- Support for the plan, in writing, emphasizing management's commitment to the program.

- Allocation of human resources either on a full or part time basis. Regardless of the time involved, these resources could be utilized in other areas.

- Funding of the plan both in the preparation stages as well as the ongoing implementation and maintenance phases.

- Awareness of the need for a recovery plan and ultimate expectations that a plan will be put in place to protect the business.

Consequently this cannot be treated lightly. With all this visibility, a lot is riding on the final implementation. This is not to imply that the job is done when implementation is done. Remember that this program is ongoing and the plan takes on a life of its own. It becomes a living, breathing document.

Schedule the Phases

Schedule review points along the way. This will be with the team and periodically with senior managers to keep them apprised. These reviews with the team will highlight progress that is being made, or in some cases what specific areas are behind schedule.

For management these review points can be either informational briefings on progress or opportunities to break down barriers in the process. A call from a key executive within the organization to a noncooperative department manager will open many doors quickly. Use this forum to keep things moving; do not hide problems.

Ensure that all tasks are delegated to a responsible person. Never try to accomplish the entire project alone. Too many details need to be addressed, and one set of eyes cannot tend to them all.

For every key contributor in the planning and implementation phase, try to assign an alternate. This is contingency: in case one person cannot get to a task, an alternate might. Brief everyone regularly. "Cross-pollinate" every aspect of the plan among multiple key players. In this way any change in staffing or job responsibilities can be readily backed up. This is an opportunity to get to a "win–win" situation.

Conduct a Final Review

One last review of what has been written should be conducted before full-scale implementation begins. A considerable amount of time has probably gone by since the inception of the planning process. Many things may have changed. Some of the changes might include the following.

- The company's business posture
- Strategic directions
- Competitive position
- Key personnel changes
- Budgetary constraints
- Systems and technologies
- Critical applications
- New products or services

Do not ignore the need to run one final check with everyone to ensure consistency in the plan. Any changes should be incorporated immediately or at least scheduled for inclusion within a specific time. This depends on the degree of change and the critical nature of the change. Review the flow of information within the organization, to assure that nothing has fallen out of a loop, remaining undone. If any key member has been kept out of the informational loop, solve the problem immediately.

Never take anything for granted. This is a business survivability issue; check everything. Then recheck the entire flow.

Order Equipment and Facilities

Assuming the final review goes well, the next step is to order products and services in a smooth flow. Wherever possible, determine the lead times required for installing and purchasing components, wiring, etc. Using the longest lead time items, begin the purchasing process. The purpose of this schedule is to be assured that when equipment arrives, or needs change, other appropriate services are in place to support these systems. Minimize delays wherever possible. The worst case would be to have equipment arrive while the supporting services remain incomplete. The result would be the need to find storage space for the equipment for whatever time it takes to complete the other tasks. This storage opens the risk of theft, abuse, or other damage. Furthermore, once an invoice is paid for equipment deliveries, the warranty period begins. It would be imprudent to use up the warranty time period before getting a chance to install and test the equipment or services first.

Rethink everything and requalify the vendors who will be providing the products or services. Make sure their posture has not diminished so that they can still provide the required support for both the short and long term.

Protect the Environment

As a means of implementation, provide as much preventative measures as possible in advance. Use the various techniques that have already been covered in previous chapters. In Fig. 9.1 a checklist is provided to conduct a final review of the process. This includes some of the many issues that can be addressed in advance.

Plan Orchestration and Advertising

As the plan begins to fall in place and the implementation begins, a single point of contact is important. This is normally the LAN manager (who is also the project manager in most cases). Control of the project rests with this one point of contact, thereby requiring some publicity throughout the organization. A well-orchestrated implementation deserves as much visibility as possible. Therefore part of the process should include periodic "news blurbs" within the organization. This can be done in an internal newsletter or via separate correspondence to all employees.

However, most LAN managers are technocrats rather than "artsy and whimsical" advertising buffs. Therefore recruit someone from the advertising or marketing communications department to assist with the appropriate text. This exposure can go a long way in maintaining the momentum of the implementation and added support for the plan. Enough "hype" used in these newsletter articles will add to the excitement and awareness of just what the plan is all about.

Training Issues

Training becomes a very serious consideration when implementation begins. Everyone must be thoroughly trained on what is expected of them. This includes the following.

1. Check all security systems, are all doors kept locked?
   ❑ Yes   ❑ No

2. Is access controlled to only those who have a need?
   ❑ Yes   ❑ No

3. Are server rooms secured?
   ❑ Yes   ❑ No

4. What form of security is used?
   - If key locks are used, who has the keys?
   - If word or combination locks are used, who has the combination?
   - How often is combination changed?

   ❑ monthly   ❑ quarterly   ❑ annually   ❑ post personnel termination

5. Are environmental services sufficient to support the current and future equipment?
   ❑ Fire   ❑ Heating, ventilation, and air conditioning (HVAC)   ❑ Electric
   ❑ Floor space   ❑ Grounding

6. Are regular inspections and cleaning done in the area?
   ❑ Clean area   ❑ Aisles kept clear   ❑ Policies enforced

7. Is any construction planned or currently ongoing?
   ❑ Yes   ❑ No
   If yes,   ❑ internal   ❑ external

8. Have uninterruptible power supply (UPS) systems been installed?
   ❑ Yes   ❑ No
   Systems characteristics:
   ❑ Sized for growth   ❑ Type   ❑ Maintenance   ❑ Bypass available

9. Do workers rush around in an uncontrolled frenzy?
   ❑ Yes   ❑ No
   If yes, check if
   ❑ Equipment is bolted down   ❑ Rushing around is limited

10. Is there evidence of poor cleaning habits?
    ❑ Yes   ❑ No
    If yes, what evidence?
    ❑ Ashes   ❑ Dust   ❑ Trash accumulating   ❑ Boxes piled high

11. Do workers show evidence that they know what they are doing?
    ❑ Yes   ❑ No   ❑ Retraining necessary

12. Are modem communications secured?
    ❑ Yes   ❑ No
    If yes, what is secured?
    ❑ Dial-in   ❑ Dial-out   ❑ Dial-back

Figure 9.1 Final checklist to protect systems.

- Trial runs of the recovery process
- Development of a training program based on who should be trained
- Cross-training for recovery team members, both primary and alternate
- Specialty training for the unique systems on-site

Training Plan

A formalized training plan should be developed that allows for the initial and ongoing needs. In Table 9.1 the types of training plans that show the various needs of an organization are highlighted. Although this list may seem extensive, the various types of training are different based on individual needs. Not everyone must be trained on specific recovery processes, but all employees need minimal amounts of training so that they know what to do, when, and why. Regardless of how much training is conducted, or how long the sessions are, the reality of the situation is "the more training given, the better the odds of a smooth recovery."

TABLE 9.1   Types of Training

Group	Training	Length
Core recovery teams	Damage assessment, LAN operating systems, application recovery First aid Help desk	Two weeks
Specialty teams	Bomb search First aid Cabling systems Electrical systems or UPS HVAC systems Fire detection or suppression Telecommunications network recovery Damage control Mitigating losses Help desk	Four to six weeks
Management	Assessment Command or control center Working with bankers, customers, media	One day
All employees	Recognizing problems Reporting problems Dealing with fires and floods First aid Relocation policy Piracy policy	One to two hours

Training Materials

After each set of training is conducted, some form of a ready reference card should be handed out. In general, all employees should know the company policies and procedures. Second, they should know what to do in the event of some out of the ordinary occurrences. A sheet ($8\frac{1}{2} \times 11$ in) of instructions folded in thirds that can be placed easily under a terminal, telephone, or desk pad may prove sufficient, depending on the organization's needs. This information sheet may be as simple as listing what to do and who to call. An example is shown in Fig. 9.2. This particular sheet of paper folded in thirds, printed front and back, can yield six separate instruction sets (kept simple) of what to do and who to call in the event of a problem. Since this sheet is folded to approxi-

IN CASE OF FIRE	IN CASE OF MEDICAL ALERT	IN CASE OF SUSPECTED SECURITY PROBLEM
1. _____	1. _____	1. _____
2. _____	2. _____	2. _____
3. _____	3. _____	3. _____
4. _____	4. _____	4. _____
5. _____	5. _____	5. _____
Safety/Security-9111	Nurse-6877	Security-9111

Front

WHAT TO DO IF SYSTEM FAILS	WHAT TO DO IF STRANGE THINGS APPEAR ON YOUR SCREEN	WHO TO CALL FOR HELP
1. _____	1. _____	1. LAN Help Desk
2. _____	2. _____	2. LANMgr.
3. _____	3. _____	3. Dept. Coord.
4. _____	4. _____	4. Training
5. _____	5. _____	5. _____
Help Desk-4357	Help Desk 4357	

Rear

Figure 9.2 A sheet of paper folded in thirds produces six columns of valuable information for all employees.

mately 3.6×8 in per panel, a lot of information can be provided. The compactness makes it convenient to slip under a telephone set, a blotter or calendar, or a terminal. It is always a ready reference a hand's reach away.

The cost to produce these materials is not very high, since this is standard size paper. Printed front and back and ready to fold, it is also simple to use this low-budget item while keeping the option open to update these sheets frequently. When lost or dirty, the cost of replacement is approximately 2¢ per sheet.

When providing the specialty and core teams with materials, the use of a bound (three-ring binder, spiral binder, etc.) action plan is more appropriate. The three-ring binder is usually a good idea, so that additions, deletions, and changes can easily be extracted and inserted into the action plan. Keep in mind that changes will occur on a regular basis; therefore the booklet must be accommodating.

Another good material to supply these teams is a laminated phone list of critical numbers. This may be the entire calling tree for the individual team or an abbreviated list. The laminated card should be sized so that it will easily fit into a wallet or purse without being obtrusive. In Fig. 9.3 a sample of a card is shown. In this particular example, the size is $2 \times 3\frac{1}{4}$ in so that it will slip into a wallet. This is also the size of a standard credit card, so it will fit in most wallets, pass cases, etc., without becoming a burden.

After the reference material, the form and format of the actual training are important. The materials to be used should include visuals of some sort, such as the following:

- Slides (35 mm)

- Transparencies

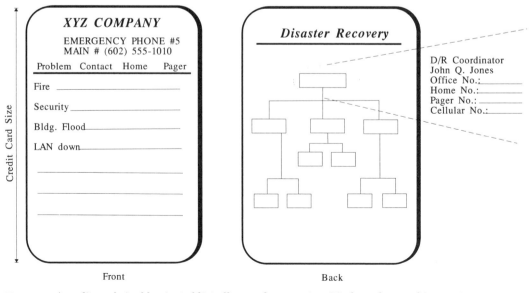

Figure 9.3 A credit-card-sized laminated list offers ready access to critical numbers and is easy to store.

- Video
- Poster or easel boards
- Computer-based training, multimedia etc.

A very useful tool in this particular case is the use of a prepackaged video or a multimedia presentation. Since most organizations will have many employees and teams to be trained, either of these tools provides for consistent delivery of the information, at the same speed and with the same enthusiasm at all times. This precludes any inconsistencies in delivery due to the presenter, time of day, rushing to get done, etc.

Additionally, the use of a videotape of multimedia on a PC (or a LAN?) will allow for new employees to get the same training as everyone else. Refresher training for those who feel they need it becomes far easier too. Scheduling problems tend to go away since most people can watch a well-produced video or multimedia presentation without tying down a presenter on a one-to-one basis.

Testing the Effectiveness of Training

After training is completed, a series of tests to verify the validity, appropriateness, and the retention level of participants should be conducted. Everyone knows that retention levels vary among individuals, but a repetitive approach, mixed with visual and audio presentations, increases the retention value exponentially. Consequently the need to repeat the individual training is required.

Many organizations fail on this part. They go through great pains to write an elaborate plan, provide the necessary backup or redundancy in systems, and spend innumerable funds on training all employees. Then, when this is completed, they feel the job is done. They never test individual scenarios to ensure that the employees grasped what they were supposed to know. This creates a possible risk to life and limb when considering the long-term consequences of this problem.

Thus the training level must be evaluated. In the event an employee has forgotten some or all of the information, do not belittle them. Moreover, do not use refresher training as a form of punishment, which is a sure road to disaster. What is needed is a simple training procedure that people can relate to and remember.

The purpose of the follow-up is to verify that the training was effective and that the employees, team members, and management all know what is expected of them. Any deviations from the expected results of the follow-up requires that the training must be updated or changed. Some of the reasons for updating the materials are shown in Table 9.2.

The Implementation and Training Plan

Eighty percent of all plans that have ever been written have never been tested. Strange as this may seem, the planning and implementation process was so grueling that when the plan was completed everyone involved wanted to

TABLE 9.2   Reasons for Updating Training Materials or Providing Refreshers

Reason	Frequency
Changes in policies or procedures within the organization	As they occur
Updates in technologies	Immediate for the recovery team; semiannual for regular employees
Changes in carrier or vendor	Monthly
Changes in personnel	Immediate
Relocation to new facilities	On occupancy
Major corporate direction changes	As needed when training is impacted
Normal follow-up shows poor results	Quarterly or semiannually
Disaster strikes; shows flaws in training	As soon as possible after recovery
Test of recovery reveal teams are confused	Immediately

get back to normalcy. It does not take very long after a plan is written and put into "production" for it to become obsolete. Imagine the time and money spent to produce a plan that will die on the shelf of someone's bookcase. It may never be opened or reviewed after it's "done."

The above scenario does not have to happen. The primary reason, as stated, is that everyone just wants to be finally done with the plan. Any project this size, whether it takes six months to two years to produce, is a monumental task. As part of the whole process, an organized schedule of events must take place. If things do not go well, a plan to overcome resistance or other barriers will help smooth the flow.

To overcome this problem, a plan to produce the plan is necessary. Use a PERT (project review and evaluation techniques) or GANTT (milestone) chart, or any other tool that feels comfortable. Keep it current, and do not be concerned with minor slippage. This is a guideline to get the job done; it is not cast in concrete to meet every task on time. The overall end date is important, but again it must be fluid. No one can accurately predict a project of six months to two years. It is a give-and-take process with each individual involved competing for time to devote to long-term planning, day-to-day tasks, and crises.

The next few paragraphs incorporate some of the steps needed to organize events and activities and structure the planning, implementation, and testing process. It is by no means the only way this can be managed; that is an individual preference.

The process for the development of the plan, implementation and so on can be broken down into several manageable phases. An example of this is shown in Fig. 9.4. This particular outline of the phases uses an eight-phase approach. Others may use ten; it depends on the individual situation. Using this information, the eight phases are then broken down into a series of tasks or subtasks. These are as detailed or general as needed by the organization.

Phase I:	Preparation and preplanning
Phase II:	Risk analysis and business impact analysis
Phase III:	Conducting a detailed inventory of equipment and critical application
Phase IV:	Backup procedures, personnel assignments, alternative locations, equipment
Phase V:	Disaster recovery organization of teams
Phase VI:	Corrective and recovery measures
Phase VII:	Documentation policies and procedures
Phase VIII:	Testing, implementation, maintenance, and training

Figure 9.4 List of the eight phases in developing the plan.

In Table 9.3 the first of the phases, preparation and preplanning is shown, by tasks. At this point the need for a support group becomes evident since one person cannot do this alone. There are just too many activities, and the plan must be integrated into other departments. No single person can know everything that goes on within an organization.

The resource requirements and start and finish dates would be determined based on the following:

1. The size of the organization

2. Amount of resources available

3. Timing based on other workloads

In phase II, the risk analysis, a detailed look at just what exposures exist within the organization, what the company posture is, and how the business would be impacted by a disaster is performed. Table 9.4 highlights some of the areas that would be considered during this phase. At this point the resources should be available to start the analysis of what could go wrong and what it would cost. Remember that in earlier chapters the cost per hour of downtime in a LAN environment averages approximately $20 to $50,000 depending on the organization and the nature of the applications running on

TABLE 9.3 Phase I Activities

Activity	Description	Resources required	Start	Finish
Introduction	Develop a case or white paper to management			
	Obtain management commitment			
	Prepare policy letter on recovery for LANS			
	Make initial presentation to management			

TABLE 9.3   (Continued)   Phase I Activities

Activity	Description	Resources required	Start	Finish
Preparation and pre- planning	Recruit team members from department heads			
	Establish goals to be met with teams and department heads			
	Research the organi- zation for other plans or master document			
	Develop job descrip- tions for team members			
	Estimate timing of project			
Preplanning for sce- narios	Determine what others have done			
	Establish if any legal requirements exist, by industry, by local ordinance, etc.			
	How long could the business survive without systems?			

the LAN. In the procedure listed in Table 9.4, the use of outside agencies such as external auditors, insurers, and actuaries would be helpful in assessing some of the values and probabilities. Use whatever resources are available.

In phase III, the inventory, details are pulled together from the myriad of components involved in running a LAN. This becomes the database of what exists at a snapshot in time. From this database all additions, deletions, and changes must be appropriately entered. This is a "moving target" that must be reviewed regularly. Furthermore, as new applications are brought onto the LAN, the licensing, interworking relationships of applications already on the LAN, and user population can and must be inventoried. Table 9.5 summarizes some of the main categories that are associated with this phase of the planning process and the structure for the final implementation. This information will take a good amount of the total time allocated to the project. In general, this point is where many users get bogged down and cannot get beyond this phase. Do not let the physical count be a problem. Use others wherever possible, such as security, safety, and facilities personnel. Do not forget to use vendors and carriers where possible.

In phase IV, the backup scenario, the project now turns to the prevention mode wherever possible. The location for recovery, if an alternative site or a

TABLE 9.4 **Phase II Activities**

Activity	Description	Resources required	Start	Finish
Risk analysis	Conduct a business impact analysis			
	Review all exposure points on the LAN			
	Determine probabilities of disaster			
	Assess communications risks			
	Outline physical security risks			
	Conduct "what if" scenarios			
	Determine worst case			
	Estimate minimum percentages of services			

hot site will be used, is reviewed. Personnel requirements during the backup and recovery process will be highlighted. In Table 9.6 the primary tasks are highlighted. This particular section is easier to address since the risks have been identified, the inventory database has been built, and the critical applications have been noted.

In phase V, organizing for disaster recovery, the plan begins to cover other issues. The first is looking at other needs (people's needs) that must be provided. Here is where the connections with other plans are used. In general, many different teams must work hand in hand to assure a smooth return to normalcy. A list of all the teams that could be involved, coupled with input from these teams, is covered in Table 9.7. The first column deals with the teams, whereas the second column addresses a series of other issues to meet human and business functionality. Again, remember that all these may appear in some other plan, so long as a reference is made to that plan. *These two columns are not interrelated but merely contain issues that are covered.*

The lists in Table 9.7 could go on ad infinitum, but the idea should be enough to portray that all things must be addressed, even if not specifically by the LAN disaster recovery plan.

In phase VI, corrective and recovery measures, the plan will address the recovery action items. These action items will include such things as what to do; when it should be done; and by whom. In Table 9.8 this is done through the logical corrective steps to get back into business as quickly as possible. Some of the action plans are written by others. Additional action plans are written by the organization but performed by others. This becomes a management problem. The disaster recovery process deals with four phases.

TABLE 9.5 Phase III Activities

Activity	Description	Resources required	Start	Finish
Inventory	Inventory all hardware; build database			
	Inventory all physical facilities to include rooms, electric, HVAC, UPS, fire detection, alarms			
	Create schematics of all cabling systems from device to network			
	Inventory all applications; verify licensing; keep records			
	Validate all communication line access to bridges, routers, gateways, modems			
	Keep records of all moves, additions, and deletions from this point forward			

These are referred to in this book as the four Rs.

The four Rs refers to the actual phases people go through and some estimate of the timing involved. These are as follows.

Item	Description	Timing
Recognize	Recognize that a problem exists.	0 to 2 h
React	Response is necessary to mitigate damages where possible. Sound the alarm and get everyone moving.	0 to 4 h
Recover	Get back into business as soon as possible. This involves a degraded amount of service as defined in the plan.	4 to 72 h
Restore	Rebuild, remodel, or do whatever is needed to put everything back to the way it was.	72 h to end

Using the four Rs, this next phase of the plan actually provides for the recovery.

As phase VII, documentation, policies, and procedures, begins, the primary concern is that many organizations have a tendency to put this phase off until later. Unfortunately what happens is that the documentation never gets done. The actual policy and procedure that started this whole process get pushed aside.

TABLE 9.6   Phase IV Activities

Activity	Description	Resources required	Start	Finish
Backup policies	Define backup and redundancy strategies			
	Assign responsibilities to LAN or department managers for backup			
	Determine frequency and type of backup			
	Create follow-up procedures to assure backups get done			
	Select equipment needed and software systems to catalog, etc.			
	Establish priority times and lists for recovery			
	Define enforcement procedures for backup			
Personnel	Train users on backup procedures			
	Establish guidelines			
	Establish frequency schedules			
	Formulate rotation procedures			
Location	Determine location for alternative or hot site			
	Negotiate contracts			
	Determine space and equipment needs			
	Locate away from downtown area			
	Establish priority of recovery and system			
	Set schedules for tests			
	Agree to responsibilities of hot site personnel			

In Fig. 9.5 the format to use is given rather than targets, as were listed in previous tables. Many LAN managers have attempted to document everything after the fact. Throughout each phase, the documentation should be ongoing. The final wrap-up should be more anticlimactic than anything else. If all the teams were putting their ideas, suggestions, and findings on paper throughout the other phases, this piece should be done in advance. It is used as a reference point in the planning and implementation

TABLE 9.7   Phase V Issues and Organizational Input

Teams	Issues
Overall team coordinator	Phone list of all employees
User department	Phone list of carriers and vendors
Information systems	Phone and contact list for clients
Telecommunications	List of bankers and insurance company
Hardware	First aid kits
Software	Disaster kit (water, crowbar, etc.)
Administrative services	Supplies for two weeks of work activities
Recovery	Record of off-site storage
Audit	Contracts
Insurance	Contracts and agreements of employees assisting in the recovery process
Safety or security	Housing
Human resources	Ground or air transport
Corporate communication (media)	Finance—petty cash
Senior management steering	
Finance or Controller	

of a recovery plan, only to remind the departments and teams that the system must be fully documented.

Finally when phase VIII, implementation, testing, training, and maintenance, arrives, the integration of the plan into the day-to-day operation of the organization should be relatively straightforward. New services and equipment can be ordered and brought into the environment easily. As the installation takes place on the research and development (R&D) LAN (or testbed), the rest of the network is isolated from any problems that may occur. All changes to policies and procedures are finalized so the organization posture with respect to the disaster recovery plan is clearly defined. No one has any doubts as to what is required.

In Table 9.9 these final activities will bring about the culmination of six months' to two years' work. But as already stated, it is never done. The fol-

1. An ongoing process.
2. Should have review and approval at several points along the way.
3. When constructive criticism from management or others is received, tend to it quickly. Do not procrastinate.
4. After every change is made, have it reviewed as a sanity check.
5. The final documentation should be anticlimactic, and the plan should be self-supporting.
6. Check it, check it, and then check it again!

Figure 9.5   Documentation phase.

low-up testing, documentation, and maintenance of the plan becomes the ongoing process to keep the plan current and all employees, team members, and management up to date and thoroughly trained. No single phase will be more important than another, but this phase is farther reaching in the longevity and perpetuity of the plan. Treat this phase very seriously; do not lose interest at the last minute.

Although the activities may appear complex throughout the eight phases, the techniques are proven. Any plan can work if it is logically thought through and has management support and a winning attitude from the team players. Build as much team spirit as possible. Keep everyone informed, and let them contribute. This can become a "win–win" situation despite the odds against it.

TABLE 9.8   Phase VI Activities

Activity	Description	Resources required	Start	Finish
Corrective measures	Fix what can be fixed before it leads to a disaster			
	Develop the action plan on how to recover: What to do When it should be done Who does it			
	Write down details of all equipment restart procedures			
Recovery	Use four Rs to define action plans			
	Define timing for critical applications			
	Recognize problem			
	React—minimize damage where possible			
	Define reporting procedures—who to call			
	Have security enforced to minimize consequential damages, after the fact			
	Notify teams—plan for assembly locations			
	When disaster is declared: Who will declare it? What is the next step?			

TABLE 9.9   Phase VIII Activities

Activity	Description	Resources required	Start	Finish
Implementation	Order all services: backup, redundant, and primary			
	Have cabling and environmental equipment installed			
	Test: acceptance period lasts for 30 days on R&D LAN			
	Update policies and lists, etc., as needed			
Training	Define training strategies			
	Develop method of delivery			
	Conduct training for: Recovery teams Specialty teams Management All employees			
	Define refresher training			
	Select new hire training strategy			
Testing	Define testing strategy			
	Establish frequency			
	Plan scheduled and unscheduled tests			
	Schedule updates after results of test			
Maintenance	Plan for periodic updates			
	Establish frequency for total updates			
	Continue ongoing training; incorporate testing results			
	Investigate new technologies and techniques			
	Distribution of copies			
	Set control procedures			
	Perform a routine check of plans; issue for accuracy check			

10

Testing and Maintaining the Plan

The plan is now developed and implemented. The major hurdles are finished; the rest is much easier. The only way to know for sure that all this effort will assist in the recovery after a disaster strikes is to test it before a disaster. How does this get done with any ease or degree of confidence?

If the testing decision is to make everything easy, no one will ever know for sure that the orchestrated set of procedures will really work. The easier the tests are, the fewer the results to indicate how valid the plan is.

Testing Strategy

A true testing strategy takes into consideration that nothing will go according to plan. In order to get a true feeling of confidence that the plan will work given any scenario is to test every possibility. However, this may be impractical since the scenarios could involve fire, flood, loss of life, hackers, etc. One clearly would not want to subject the organization to these real-life situations unless it was absolutely necessary as a result of a true disaster. Therefore a strategy must be designed that will use as much realism as possible in a simulated role. In Fig. 10.1 a series of questions must be addressed while defining this portion of the plan. The questions are geared to the real business issues that focus on what to test and to what degree, and the costs associated with testing.

The strategy chosen may be to test some of the simpler portions of the disaster scenario. A simulated fire could be used. In a simulation, even though a real fire does not exist, a test could put the teams through all the motions as though a fire really existed. This could be a check on the readiness of the teams. Their knowledge of what should be done is certainly worth an extensive evaluation.

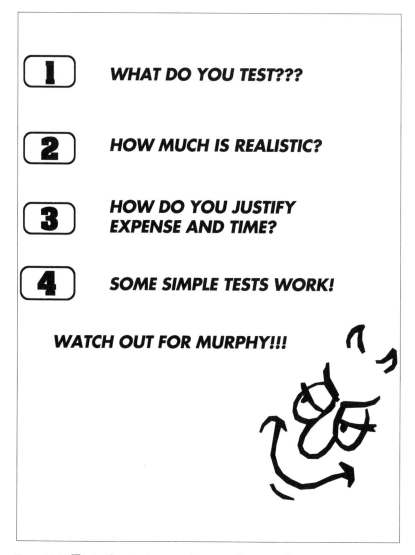

Figure 10.1   The testing strategy must answer these questions.

In general, after the plan is in place, some simple tests might be conducted. These will give the teams the opportunity to put their best foot forward and have some success at running through the procedures. If the first test they had was very difficult, their failure (and they will fail) could destroy their confidence and support. They will be so demoralized that they may just give up. Having come this far in the process, they need a small reward. They need to see it succeed.

Some simple power outages could be arranged by scheduling a test of the uninterruptible power supply (UPS) system. Or a cable could be disconnected, simulating a cut cable, to see if a rewiring or rerouting scenario would work as planned. These tests should be done on non-mission-critical devices, because as the saying goes, "Murphy is alive and well": whatever can go wrong will go wrong! The power tests might well be done during off-hours so that any glitches will not affect the work force. As may be obvious through this discussion, the testing must begin slowly and then made more difficult as time goes on.

Who Should Be Involved in Testing the Plan?

Selecting the appropriate team members to participate in a test is also essential. Many teams exist, each with their own function and role. By choosing who to test as well as what to test, credibility can be achieved. What people should be involved?

The participants could include the teams shown in Fig. 10.2, provided in a checklist format of an all-inclusive test, using everyone on the list.

How Often Should Tests Be Run?

With all the issues that have been covered, the frequency of a full test should be at least once a year. However, in reality the tests should be conducted as

Figure 10.2 A checklist of teams that can be involved in a test.

often as necessary to keep everyone sharp and fully trained. As team members change, the new members will have to be indoctrinated as quickly as possible. After an initial training period, a test is in order. Figure 10.3 is a checklist of the choices for frequencies of tests.

Use any real disruptions to the network as an opportunity to conduct a test. The statistics covered earlier in this book could indicate the opportunity to test pieces of the plan twice a month. Learn from every disruption. Did things get back into operation within a reasonable amount of time, or was an excessive recovery time evidenced? This would indicate that a problem could exist. Find out what went right and wrong. Then modify the plan to overcome the problems. Often the people involved in real day to day problems become complacent or apathetic. If this is just another network failure like all the others, they tend to handle the failures as a matter of course, without urgency. This attitude can be dangerous and disruptive if a real disaster strikes. Get to the root of this attitudinal handicap and solve it now! When disruptions to the LAN occur, keep a record of what the causes were. Use Fig. 10.4 as a checklist to maintain a historical file, then learn from the results.

Figure 10.3 Checklist for test frequency.

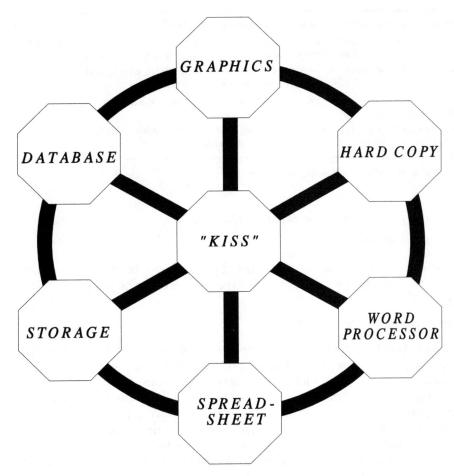

Figure 10.4 Use the tools to integrate the process.

Plan Modifications

Rewrites of the plan may be necessary after the results of a test are evaluated. Use the KISS method (keep it short and simple) to allow the teams to perform quickly and efficiently. Do not write the plan to be the world's greatest literary work. Keep it functional. The use of several packages were mentioned before as tools to keep the plan current. This included the items listed in Fig. 10.5, which shows the interrelationships of each of the packages.

Rewrites to the testing procedures and the plan will be required whenever

- New hardware is installed
- New applications are introduced
- Procedural changes are instituted
- Personnel changes occur

1. The word processor is the tool to write and edit the necessary words for the team to follow.
2. A relational database holds the inventory information. Since the same components may be referenced in more than one place, the relational database takes care of the updates in various locations.
3. A spreadsheet is useful in projecting the necessary budgetary expenses. With a tie-in to the word processor and the database, the spreadsheet program becomes a powerful tool.
4. Graphics are tools to create simple flowcharts and pictures. By using easy to follow charts, the recovery process can be expedited.
5. A storage system on a PC (or the LAN) is used to keep all these files available and handy. It must be backed up as much as any other data on the LAN, if not more so.

Figure 10.5 Tools used to update plans and testing procedures.

- Relocation of equipment or departments occur
- Wiring changes occur

By conducting these tests the above-noted changes can be detected to allow for quick rewrites to the action plan and the implementation plan.

Using Disruptive or Nondisruptive Tests

A full-scale test that requires the complete shutdown of a LAN (or multiple LANs) and possibly the relocation to another facility can only be viewed as disruptive. This is the toughest test to accomplish since critical information will be impacted, personnel will be inconvenienced, and management will be generally inclined to be less than supportive. However, this is a necessary drill at least once a year or perhaps once every two years at the worst case.

As a compromise, a nondisruptive version can be considered. This may include a scenario of using only the teams who will be involved with moving off-site. This has been done for years by data processing departments while the rest of the business functioned normally. Although it only tests portions of the process, it does at least keep the readiness level high. Figure 10.6 is a comparison of disruptive and nondisruptive tests.

DISRUPTIVE	VS.	NONDISRUPTIVE
DOLLAR IMPACTS HIGH		LESS IMPACT
MANAGEMENT FROWNS ON IT		MANAGEMENT CONDONES
TRUE TEST		PARTIAL RESULTS
PERSONNEL INCONVENIENCE		NO ONE NOTICES

Figure 10.6 A comparison of the tests in a disruptive vs. nondisruptive mode.

A test should be as tough as possible, testing, testing, and testing until it fails! Only through a failure can a weak link be discovered and prove that the test was significant. It is worthless to conduct a dry run to show management how good it works; the true results come when an area that needs shoring up is uncovered, although this is what usually happens. The team or the LAN manager wants to show positive results for the time and investments that have been made in the planning process. To show the results they set the test to succeed 100 percent. They can therefore go to management and show them the value of the investments and the plan. Unfortunately, this is the old "smoke and mirrors" routine. Of course the test succeeded, since it was set up to do so. However, a better test would have been to prepare management for the inevitable failures of small parts of the plan. This shows management that the tests are real and have a purpose.

Therefore the scenario selection must vary with each test. An example might be to test one time using fire as the disaster, and the next with a major virus infecting the system. Keep everyone alert. Test different pieces and different people on various optional scenarios. Test parts of the plan regularly.

Once again this requires management support and commitment to enforce the procedures. Gaining that approval must be built into the plan or else the whole process may fall apart. Many companies have been known to cite management's reaction for a planned test. The results are less than heartening.

Vary the Tests

Nothing in the book of rules says that all tests must be equal. As a matter of fact, the best results will come when the tests are varied. The results will truly reflect the ability to recover the LAN when a mix of tests are involved. Some of the variances might include the types as shown in Fig. 10.7. By using this format, the following scenarios could be used:

1. Use announced preplanned dates such as the last Friday of June every year. People tend to clean up their acts in preparation for such a test. This is fine if it accomplishes some positive results.

2. Use unannounced (surprise) tests to see how quickly teams muster and react. This could be on a Friday afternoon before leaving for a long weekend. Or an option might be a call to the team members' homes at 5:00 A.M. or 5:00 P.M.

3. Despite the rules, introduce some complexities spontaneously. "Kill" someone off and see who jumps in to help out (figuratively speaking only). Have a management representative make the call (announcement) instead of the LAN manager. The teams will not know if the announcement is real or not!

4. Run a backup test. Take your on-line systems off-line and recover from your backup systems and tapes. This ensures that the backups are as current as the actual on-line files and equipment revisions. This also assures that the tapes are good. Too many times, people have written data to tapes and then could not recover data from the tapes because of bad read/write heads or incompatible systems.

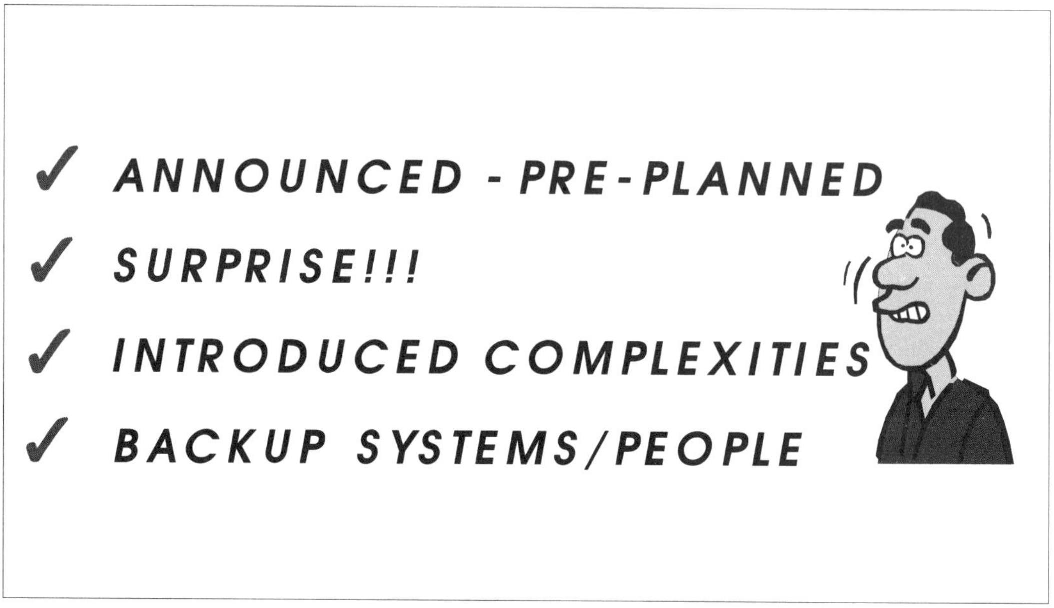

Figure 10.7   Vary the tests to get true results.

Furthermore, use the backup members of the team, while the primary member observes. See things through different sets of eyes.

These actions will improve the results and the dynamics of a true plan. Remember,

Document it, document it, document it!

Then...test it, test it, and test it until it fails!

Plan Maintenance

As the plan evolves into a usable document it must then become an ongoing entity. Writing and implementing a plan that never gets updated is both a waste of time and money. As mentioned earlier in this book, 80 percent of all the plans that have been written have never been tested. Most of these (70 to 80 percent) were the result of responding to an audit report. Once the plan was hammered together, the departments felt that the auditors would be satisfied.

Since the plan has never been tested, one could assume that it is still in its original form. Given a two year cycle in today's business climate and the way technologies change, one could assume that it is therefore obsolete. Can a company truly recover if their plan is obsolete? Sure they can, but at a greater risk of cost, image, and credibility. A Fortune 500 company may not cease to exist because their plan is obsolete. It will just throw more dollars and people at the problem until it gets fixed.

Unfortunately, many small to midsize companies don't have the luxury of money and people. They could suffer immeasurable consequences after a disaster. The statistics are full of companies that went into bankruptcy after a disaster struck. Many of them were able to get through the disaster, but it took them much longer than it should have. As a result, their finances were crippled, their customers began to seek out new suppliers to reduce the dependency on the company, and their employees lost confidence in the stability of the company's ability to exist. All of these factors combined to drive these organizations out of business. Is this what is in store for other firms that either do not have a plan or do not put the effort into maintaining it?

Frequency of Updates

How often should the plan be revised? As stated several times, it depends on what transpires within an organization. At a minimum the plan should be reviewed at least once a year. If no changes are necessary (which is highly unlikely), then a general distribution memo should be sent out stating so.

However, Fig. 10.8 summarizes when updates are necessary. Use this as a form of follow-up to ensure constant review of this all-important document is afforded.

Degree of Change

Not every scenario noted in Fig. 10.8 requires a complete rewrite of the plan. Perhaps a few sections or pages may require updating. So be it. At least once a month, the changes could be consolidated and sent out to the appropriate individuals. Use some form of bright border or banner when sending out changes. When these documents arrive on an individual's desk, they will

---

- New technologies are installed
- New applications are implemented
- Additional organizations are merged or acquired
- Team members changed
- Vendors and/or carriers changed
- After a test is completed
- New wiring topologies are used
- A move takes place
- Changes in business direction occur
- Competitive position increases or decreases
- New management takes over
- Phone numbers (home, office, pager, cellular) changed
- Budget constraints are imposed that might limit performance
- Other differences not listed are overlaid into the plan

Figure 10.8   When to update the plan.

catch everyone's attention and demand immediate action. If the changes are frequent, the bright bordered pages will outnumber the original pages. Then it is time to send out a whole new set of documents.

Control

To control the changes being put into the plan, there are several alternatives available. The degree of complexity is different depending on the choice selected. The choices are contained in Table 10.1 along with the advantages and disadvantages listing the choice of each.

As can be seen from the options, there is no one easy answer. The books must be treated confidentially, since they contain a lot of numbers (home, pager, etc.) of key employees and vendors alike. Therefore some form of control is necessary.

Distribution

The question always arises: Who should get copies and how many? This is always a tough issue to deal with. However, a rule of thumb is that minimal-

TABLE 10.1   Control of Changes

Choice of distribution	Advantages	Disadvantages
1. Send out pages, let users handle them.	*a.* Easy to distribute. *b.* No follow-up required. *c.* Inexpensive.	*a.* Changes may never get into book. *b.* User may just stuff changes in front of book. Result: outdated.
2. Send out pages with bright borders with instructions on where to put change pages	*a.* Easy to distribute. *b.* No follow up. *c.* Not real expensive.	*a.* Same as disadvantage 1*a* above. *b.* More expensive with no effective yield.
3. Send out controlled sequence pages. Users must remove old pages, insert new ones and return old ones.	*a.* Good control. *b.* Changes are accounted for. *c.* Administration on a control number basis. *d.* Easy to follow-up with users who do not return pages.	*a.* Very expensive. *b.* Very heavy overhead. *c.* Administrative nightmare. *d.* Becomes a burden.
4. Notify users to return books to disaster recovery coordinator. The coordinator removes old pages, inserts new pages, and returns book to owner.	*a.* Most control. *b.* Changes are verified. *c.* Administration control on a named basis. *d.* Easy to follow-up.	*a.* Most expensive. *b.* Very heavy overhead. *c.* Frequent changes result in risk of lost books. *d.* Administrative nightmare. *e.* Burdensome. *f.* Disaster could strike while books are in transit.

ly any person listed in the plan and having a responsibility to assist in recovery should get a copy of the action plan applicable to them. This could include vendors and carriers, so the need for confidentiality is magnified.

For all team members who have recovery actions to perform, a minimum of two copies should be issued. One is kept in the office, available during business hours, stored under lock and key. A second copy is kept at the team member's home in case a disaster occurs during off-hours. This allows them to spring into action immediately without having to chase after information.

Critical team members who are asked to be available (or needed) at all times have been known to have three copies. The third is kept in a locked box in the trunk of their car. This way if they are not in the office or at home, they still have the information readily available to them. Many use the third copy in their briefcase, so that while traveling out of town, they have the information and can contribute immediately. Each organization has different guidelines that will provide direction.

The importance of the plan being current has been stressed, but can never be stressed enough. Once the plan is in place, it will constantly need review. Using kickoff points and the appropriate distribution system and the testing procedures could make the plan easier to support.

Management has committed a significant amount of funding and resources to get to this point. Therefore the plan must be kept current in support of their commitment. Remember it is the perpetuity of the business that is at stake here.

The plan is now ready to go into full maintenance mode; the organization should be prepared to react and recover in the event of a disaster with the LAN. Good luck.

Vendor Products
and Services

Vendor name and address	Product type	Product name
ATS 1160 Ridder Park Drive San Jose, CA 95131 Phone (800) 359-3580	Hardware mass storage products	Various products
AT&T Computer Systems PO Box 25000 Greensboro, NC 27301 Phone (919) 279-3024	Hot and cold sites, consulting services	
Absolute Security 63 Great Road Maynard, MA 01754 Phone (508) 897-1991	Audit system, file analysis	LANaccess LAN Investigator Plus
Advanced Digital Information Corp. 14737 NE 87th Street Redmont, WA 98073 Phone (206) 881-8004	Backup systems	DAT Autochanger
Advanced Information Management 12940 Harbor Drive Woodbridge, VA 22191 Phone (703) 643-1002	Disaster recovery planning software	AIM/Safe 2000, AIM/Bank 2000
Agway Data Services, Inc. PO Box 4862 Syracuse, NY 13221-4862 Phone (315) 477-6510	Hot sites	
American Power Corp. 350 Columbia Street Peace Dale, RI 02883 Phone (401) 789-5735	Power systems	Various UPS products
Apricot 111 Granton Avenue Suite 401 Richmond Hill, Ontario Canada L4B 1L5 Phone (416) 492-2777	Servers	VX FT 486 server
Vendor name and address	Product type	Product name
Autosig Systems, Inc.	Biometric system,	Various products

PO 165050 Irving, TX 75016 Phone (214) 258-8033	signature verification	
Backup Recovery Services, Inc. 1620 NW Gage Blvd. Topeka, KS 66618 Phone (913) 232-0368	Hot sites	
Banyan Systems, Inc. 120 Flanders Road Westboro, MA 01581 Phone (508) 898-1000	Servers, operating systems	Banyan/CNS VINES
Brightwork Development Inc. 766 Shrewsbury Ave. Tinton, NJ 07724 Phone (908) 530-0440	Utilities software	LAN Automatic Inventory
Business Resumption Planners 572 Emerald Avenue San Carlos, CA 94070 Phone (415) 592-5995	Disaster recovery planning software	Analysis 2000
CACI Products Co. 3344 North Torrey Pines Court La Jolla, CA 92037 Phone (619) 457-9681	Network design software	LANNET II.5
Cable Technology Group, Inc. 55 Chapel Street Newton, MA 02160 Phone (617) 969-8552	Design and documentation systems	LAN-D/S
CHI/COR Information Management, Inc. 10 South Riverside Plaza Chicago, IL 60606 Phone (312) 454-9670	Disaster recovery planning software	TRPS, Disastar for Windows TRPS
CSC CompuServe 118 MacKenan Drive, Suite 300 Cary, NC 27511 Phone (919) 460-1234	Hot sites	
Chubb Insurance Group 15 Mountain View Road Warren, NJ 07061	Insurance	
Citadel Systems Inc. 9800 Northwest Frwy., Suite 610 Houston, TX 77092 Phone (713) 686-6400	Netware software	NetOff
Citel America, Inc. 1111 Parkcentre Blvd., Suite 474 Miami, FL 33169 Phone (800) 248-3548	Hardware surge protectors	Networth vLan, vNet Surge Protector, and various other models

Vendor name and address	Product type	Product name
Clary Corp. 320 West Clary Avenue San Gabriel, CA 91776 Phone (818) 287-6111	Power systems	ONGUARD UPS
Comdisco Disaster Recovery Services 6111 N. River Rd. Rosemont, IL 60018 Phone (708) 698-3000	Hot and cold sites, recovery software, consulting services	Various software products
Command Record Services, LTD. 195 Summerly Rd. Brampton-Ontario, CANADA L6T4P6 Phone (416) 293-1161	Off-site storage	
Command Software Systems 28990 Pacific Coast Highway Malibu, CA 90265 Phone (800) 423-9147	Workstation lock, encryption system	Freeze Languard
ComNetco Inc. 29 Olcott Sq. Bernardsville, NJ 07924 Phone (201) 953-0322	Anti-virus protection	Virusafe
Computer Solutions, Inc. 397 Park Avenue Orange, NJ 07050 Phone (201) 672-6000	Hot site	
CrossComm Corp. PO Box 699 Marlborough, MA 01752 Phone (800) 388-1200	Routers, network management, software, and hardware	Various products
Cylink 310 N. Mary Avenue Sunnyvale, CA 94086 Phone (408) 735-5800	Hardware fax security, hardware link security, voice security	SecureFX, CIDEC-LS, CIDEC-HSi, CIDEC-VHS STM-9600
DCA 1000 Alderman Drive Alpharetta, GA 30201 Phone (404) 442-4545	Servers	10Net
Data Archives, Inc. 4 Litho Road Lawrenceville, NJ 08648 Phone (609) 771-0116	Off-site storage and electronic vaulting	
Data Protection, Inc. 7558 Southland Orlando, FL 32809 Phone (305) 851-8557	Off-site storage	
Data Protection, Inc. PO Box 1966 West Chester, PA 19380 Phone (609) 771-0116	Off-site storage	

Vendor name and address	Product type	Product name
Data-Tech Business Resource Centre, Ltd. 126 Lower Richmond Road London SW15 1LN United Kingdom Phone 081 780 2412	Technical training	Disaster recovery planning for LANS, PC and LAN Security
DataTech Institute PO Box 2429 Clifton, NJ 07015	Technical training	Disaster Recovery Planning for LANS
Data Vault Corp. 139 Newbury Street Framingham, MA 01701 Phone (508) 879-5510	Off-site media and paper storage	
Delaney Recovery Services, Inc. 888 First Avenue King of Prussia, PA 19406 Phone (215) 992-1081	LAN and workstation hot site	
Deltec 2727 Kurtz Street San Diego, CA 92110 Phone (800) 854-2658	Power systems, UPS interface software	2000 Series UPS PowerCheck
Dickson Company 930 S. Westwood Avenue Addison, IL 60101 Phone (800) 323-2448	Environment monitors	Temperature and humidity recorders
Digital Dispatch Inc. 1580 Rice Creek Rd. Minneapolis, MN 55432 Phone (612) 571-7400	Antivirus protection	Data Physician
Digital Equipment Corp. 3 Results Way, MR03-2R15 Marlboro, MA 01752 Phone (508) 467-7018	Hot and cold sites	Business protection services portfolio
Disaster Recovery Services, Inc. 427 Pine Ave., Suite 201 Long Beach, CA 90802 Phone (310) 432-0559	Disaster Recovery planning software	Disaster Recovery 2000
Ecco Industries 130 Centre Street Danvers, MA 01933 Phone (508) 777-7750	Biometric system	VoiceKey
ECCS One Sheila Drive Building 6A Tinton Falls, NJ 07724 Phone (800) 322-7462	Backup and storage systems, RAID	DFT-1, FFT-1, DFT-5, FFT-5
EyeDentify PO Box 3827 Portland, OR 97208 Phone (503) 645-6666	Biometric system	EyeNet

Vendor name and address	Product type	Product name
FingerMatrix 30 Virginia Road North White Plains, NY 10603 Phone (914) 428-5441	Fingerprint software and hardware	Various products
Frye Computer Systems Inc. 19 Temple Pl. Boston, MA 02111 Phone (617) 451-5400	LAN utilities	NetWare Console Commander
Gateway Technologies 180 West Church Road King of Prussia, PA 19406 Phone (215) 354-0330	Workstation security software	Privacy
General Software PO Box 2571 Redmond, WA 98073 Phone (206) 391-4285	Protocol and network analyzer software	The Snooper, EtherProbe
GigaTrend, Inc. 2234 Rutherford Road Carlsbad, CA 92008 Phone (619) 931-9122	Backup systems (with database management system)	Master 8500 Series
Globus Systems, Inc. 1447 McAllister Street San Francisco, CA 94115 Phone (800) 538-4701	Physical security software (MAC systems)	Security Force
IBM Corp. 1300 N. 17th Street 6th Floor Arlington, VA 22209 Phone (703) 841-3149	Consulting services	IBM Business Recovery Services
International Data Sciences 501 Jefferson Blvd. Warwick, RI 02886 Phone (800) 437-3282	Test equipment	Numerous products
Iomega Corp. 1821 West 4000 South Roy, UT 84067 Phone (800) 777-6654	Disk storage hardware	Various products
Iron Mountain Group 745 Atlantic Ave. Boston, MA 02111 Phone (617) 357-9034	Off-site storage	
LAN Systems 300 Park Ave. S, 15th Floor New York, NY 10012 Phone (212) 995-7700	Secure menus	LAN Shell, Reference Point
LDI Disaster Recovery Corp. 30700 Carter Street Solon, OH 44139 Telephone (216) 248-0991	Hot and cold sites	

Vendor name and address	Product type	Product name
Lattice, Inc. 3010 Wood Creek Road Suite-A Downers, IL 60515 Phone (800) 444-4309	Encryption	SecretDisk II
McAssee Associates 3350 S. Blvd. # 14 Santa Clara, CA 95054 Phone (408) 988-3832	Antivirus protection	Viruscan
Maynard Electronics, Inc. 36 Skyline Drive Lake Mary, FL 32746 Phone (407) 263-3500 Europe: Maynard Electronics, Inc. Coronation Road High Wycombe, Bucks HP12 3T United Kingdom Phone (0494) 473434	Tape backup hardware	Numerous products
Microcom, Inc. 500 River Ridge Drive Norwood, MA 02062 Phone (800) 822-8224	LAN bridges	MCBR
Micronyx 1901 N. Central Expressway Richardson, TX 75080 Phone (214) 690-0595	Audit system Software security	TriSpan SAFE
Mobile Computer Recovery 211 College Road East Princeton, NJ 08540 Phone (609) 452-8980	Mobile data centers	
Network-1, Inc. PO Box 8370 Long Island City, NY 11101 Phone (718) 932-7599	Virus software (VAX) Security software and consulting	C4V (for VAX/VMS systems) SFE
Novell, Inc. 122 E. 1700 South Provo, UT 84606 Phone (800) 638-9273	Operating system Operating systems utility	NetWare SFTIII
On Disk Software 11 Waverly Place New York, NY 10003 Phone (212) 254-3557	File analysis	Quarantine
Ontrack Computer Systems 6321 Bury Drive Eden Prairie, MN 55346 Phone (800) 752-1333	Backup systems	Disk manager

Vendor name and address	Product type	Product name
Palindrome Corp. 850 E. Diehl Road Naperville, IL 60563 Phone (708) 505-3300	Backup systems	Various products
Panda Systems 801 Wilson Road Wilmington, DE 19803 Phone (302) 764-4722	Antivirus protection	Dr. Panda Utilities
Para Systems, Inc. 1455 LeMay Drive Carrollton, TX 75007 Phone (214) 446-7363	Power systems	Minuteman Series UPS
Progressive Computing 814 Commerce Drive Oak Brook, IL 60521 Phone (708) 574-3399	Test equipment	LM1 Analyzer
Provident Recovery Systems 118 Mac Kenna Drive Suite 300 Cary, NC 27511-8507 Phone (919) 481-0011	Mobile recovery facilities	
Quaid Software 45 Charles St. E., 3rd Floor Toronto, Ontario Canada M4Y 1S2 Phone (416) 961-8243	Antivirus protection	Antidote
Recognition Systems 62 S. San Thomas Aquino Rd. Campbell, CA 95008 Phone (408) 364-6960	Biometric system	ID3D-R Handkey
Recovery Management, Inc. 435 King Street PO Box 327 Littleton, MA 01460 Phone (617) 486-8866	Disaster recovery software	Rexsys and Rex-Collect
Saber Software PO Box 9088 Dallas, TX 75209 Phone (800) 338-8754	Workstation lock Secure menus	Secure Saber Menu System
Silver Oak Systems 8209 Cedar Street Silver Springs, MD 20910 Phone (301) 585-8641	Security and antivirus software	Iron Clad 2.0
Spider System, Inc. 12 New England Executive Park Burlington, MA 01803 Phone (617) 270-3510	Network analysis	SpiderMonitor, SpiderAnalyzer

Vendor name and address	Product type	Product name
St. Paul Fire & Marine Insurance 385 Washington Street St. Paul, MN 55102 Phone (612) 221-7911	Insurance	
SunGard Planning Solutions, Inc. 1285 Drummers Lane, Suite 300 Wayne, PA 19087-1572 Phone (215) 363-2227	Consulting services	
SunGard Recovery Services 401 N. Broad Street Philadelphia, PA 19108 Phone (215) 351-1300	Hot and cold sites	
The Systems Audit Group, Inc. 25 Ellison Road Newton Center, MA 02159 Phone (617) 332-3496	Information sources	Disaster Recovery Yellow Pages
TC International Consulting, Inc. PO Box 51108 Phoenix, AZ 85076-1108 Phone (602) 759-7502	Consulting services	Disaster Recovery Planning
Tallgrass Technologies Corp. 11100 West 82nd Street Lenexa, KS 66214 Phone (913) 492-6002	Backup storage	FileSECURE 80/15/1300
Telebyte 270 E. Pulaski Road Greenlawn, NY 11740 Phone (516) 423-3232	Protocol analyzer	PC Comscope
Telecommunications Techniques Corp. 20410 Observation Drive Germantown, MD 20876 Phone (800) 638-2049	Test equipment	T-BERD, FIREBERD
Vortex Systems, Inc. Hq. 800 Vinial Street Pittsburgh, PA 15212 Phone (412) 322-7820	Virtual disk subsystems	TC 376
International Vortex Distributor: Westcon 6830 Cote De Liesse St. Laurent, Quebec, Canada H4T 2A1 Phone (514) 344-5151		
WBS and Associates 7620 Little River Tpk. Suite 600 Annandale, VA 22003-2630 Phone (703) 941-0270	Software and hardware backup	LAN SWEEP

Vendor name and address	Product type	Product name
Wang UK Limited Wang House 1000 Great West Road Brentford, Middlesex TW89HL United Kingdom Phone 44-81-231-351		
Wandel & Goltermann, Inc. 1030 Swabia Court Research Triangle Park, NC 27709 Phone (919) 941-5730	LAN protocol analyzer	DPA-10, DA-30, DA-31
Weyerhaeuser Recovery Services Park Center Two Building Tacoma, WA 98477 Phone (800) 654-9347	Hot and cold sites	

Glossary

Alternate Routing A process of using lines other than the main link for the delivery of data from network to network.

Amplifier A device for strengthening the analog signal to a level needed by other devices on the system or network.

Arrestor A device used to protect telephone equipment from lightning and electrical spikes. An arrestor is typically a carbon block or a gas tube. After lightning strikes, the gas ionizes and causes a low resistance to ground, thereby draining the surge away from the equipment.

Asynchronous Transmission A transmission in which time intervals between transmitted characters are unequal or random.

Attachment Interface Unit (AUI) The cable, connectors, and circuitry used to interconnect the physical signaling layer and MAU.

Backbone Closet The closet in a building where the backbone cable is terminated and cross-connected to the horizontal or vertical distribution cables.

Backup A copy of computer or communications data on an external storage medium, such as a floppy or a tape. Computers experience glitches, thereby causing them to lose information. Backups save time in returning to normal operations after a failure or loss of data.

Bandwidth The range of frequencies that can be carried across a medium, assigned to a channel or system. In a LAN bandwidth is usually referred to the effective throughput in bits per second (bits/s).

Baseband A single electrical input on a medium or carrier system in its original unmodulated state.

BNC A 50-ohm coaxial connector; abbreviation for bayonet Neill-Concelman.

Bridge The hardware and software necessary to connect two networks that use the same technologies to communicate. Operates at the data link layer (layer 2) or medium access control (MAC) layer of the OSI reference model.

Broadband A communications channel using a high data transmission speed. Typically used in defining systems based on cable TV technology. Many signals can be present on a single cable at the same time, as long as they are separated and operate at different frequencies.

Bug An indescribable glitch that affects proper operation of a system. Also referred to as an unwritten feature, which exists in poorly written software.

Bus   An electrical path on a circuit. The bus usually refers to a baseband technology with a single user on the cable system at a time. The bus appears to be a straight piece of cable.

Bypass   A method of using local communications other than the local-exchange carrier, either because the local telephone company is too expensive, or it cannot give you the bandwidth, routes, or service you want.

Cable   Different types of wires or groups of wires capable of carrying data and image traffic.

Cable Riser   Cable running vertically in a multistory building.

Cable Vault   The room under the main distribution frame in a central office or customer premises; a prime target for the spontaneous outbreak of fire. It is usually not protected by Halon or other suppression systems, yet the buildup of methane gas is very possible in these areas.

Call-Back Modem   A modem that is designed to call you back, at a predetermined location. A form of security to prevent dialing directly into the network; requires a password.

Carbon Block   A device used to protect cabling and systems from lightning strikes. The carbon block consists of two electrodes spaced so that any voltage above the design level is arced from line to ground. They are effective, but can be destroyed if high voltage is directly applied.

Carrier Sense   Listening to the cable system for the presence of another transmission before attempting to transmit data.

Central Processing Unit (CPU)   The brain of a general purpose computer that interprets and acts on instructions.

Cold Site   An empty shell offered by a disaster recovery company or provided by the user, into which a complete telecommunications system or computer system can be moved.

Collision   Multiple transmissions occurring on a baseband system causing the garbling of the original data.

Conduit   A piping system used to carry network and electrical cables. A conduit protects the cable and prevents burning cable from spreading flames or smoke. Most fire codes require conduits in high-rise office buildings, wherever the cable passes through human space, or plenums serving return air to human space.

CSMA/CD   Carrier sense multiple access/with collision detection. A network access method used to prevent or manage collisions of data packets.

Data Link Layer   The layer-2 function of the OSI reference model that defines the medium access control and the logical link control functions.

Dial Backup   A network scheme using dial-up telephone lines as a temporary replacement for failed leased data lines.

Disaster   The failure of a critical system, network, or power in your computer environment; the unrecoverable loss of data or the interruption of the day-to-day business functionality with no plans in place to get back into business as quickly as possible.

Diverse Routing   Having more than a single physical link between two networked points. Preferably three different paths between points A and B, depending on the economics. *Diverse routing* is the physical connection, as opposed to *alternate routing,* which is the process of using the diverse routes.

DOS   Disk operating system. The operating system for a computer using disks.

E-Mail   Electronic mail. A system used to send and receive messages between and among users of a computer network.

EMI   Electromagnetic interference. The interference caused on a network by electrical induction or crosstalk. This form of interference will appear as noise on the network, therefore causing data corruption.

Encryption   The ciphering of data (or voice) by applying an algorithm to plain text in order to convert it to cipher text, or secured text.

Ethernet   A LAN with all the related protocols developed by Digital Equipment Corporation (DEC), Intel Corporation, and Xerox Corporation. It is the most widely used and installed networking technique in the LAN arena.

Facility   In telephone terms, it is the phone or data line. In business terms it can also mean the physical plant, structures, or buildings on a campus, in a high-rise or low-rise building.

Fading   The reduction in signal intensity of one or all the components of a transmitted signal.

Frame   A unit of transmission that carries protocol and data information.

Frame Relay   A network routing technique to transmit frames across a wide area network. The frame relay systems use the bottom two layers of the OSI reference model to expedite frame delivery and reduce overhead.

Gateway   The hardware and software required to allow two different networks to communicate with each other. A gateway provides the protocol conversion from one networking architecture to another and uses all seven layers of the OSI reference model.

Ground   A problem that exists when a circuit is accidentally crossed with a grounded conductor. A wire designed to carry voltages (i.e., lightning spikes) away from electrical and electronic components that may be damaged by the surge. Improper grounding is a significant contributor to LAN system problems.

Ground Loop   Occurs when a circuit is grounded at one or more points. It can cause serious LAN system problems.

Hacker   A person who "hacks" away at a computer or system until he or she gets into the programming or features. Hackers can be destructive or merely a nuisance.

Hot Site   A site fully equipped with a telecommunications system, or computer facility, ready to operate with short-term notice. Full environmental control and systems are in place; the customer merely moves his or her people in and begins working.

Hot Standby   Backup equipment that is kept running (usually in parallel) in case some equipment fails. The hot standby then takes over, since it has been constantly updated.

IDF   An intermediate distribution frame.

IEEE   Institute for Electrical and Electronic Engineers.

Infrared   Portion of the electromagnetic spectrum used for fiber optic transmission and, for a short distance, through the air for data transmission.

Interexchange Carrier (IEC)   The long distance telephone companies that carry traffic (voice, data, and video) across their networks.

Interface   A shared boundary between elements of a system defined by common interconnections, signals, electrical characteristics, and exchange of signals.

LAN   Local area network. A short distance network used to connect terminals, computers, and peripherals under some form of standard.

LEC   Local-exchange carrier. The new term to describe the local phone company, whether a Bell Operating Company or an independent that provides local data service.

Line of Sight   A clear unobstructed line between two communications sites, typically for microwave, infrared, or other through-the-air transmission systems.

Local Loop   The physical wires that run from the customer's premises equipment to the CO. The point at which most failures occur due to construction, cuts, old cabling, flooding in the tubes, etc.

MAN   Metropolitan area network. A network that extends to approximately 50km range, operating at speeds from 100 to 150 Mbit/s.

MAU   Medium attachment unit, or multistation access unit. A unit that allows a single device or multiple devices to access a medium.

Modem   MOdulator/DEModulator; a contraction of the two functions provided by the data communications equipment. A device that transmits signals across a circuit and acts as a digital-to-analog convertor.

MTBF   Mean time between failures. The average time a manufacturer estimates before a failure occurs.

MTTR   Mean time to repair. The vendor's estimated average time to fix a problem.

Node   Any intelligent station, terminal, computer, or other device in a computer network.

Novell   A company that manufactures and distributes a network operating system (NOS). The industry leader in terms of networks running this NOS.

Operating System   A program that organizes and manages the hardware and software environment of a computer system.

Personal Computer   A microcomputer that is used for personal or individual work efforts.

Plenum Cable   Fluoropolymer-insulated and -jacketed cables with fully color coded insulated copper conductors. Also called Teflon cabling, which has been designed as a low-smoke-producing cable.

Polyvinyl Chloride (PVC)   A thermoplastic material composed of polymers of vinyl chloride. A tough water- and flame-resistant insulator.

Proactive   Taking the initiative; doing something before someone forces you to do it.

Protocol   The formal set of rules governing the timing and format of message exchange in a communications network or a LAN.

RAID   Redundant array of inexpensive disks. A redundancy technique employing one or more hard disks to prevent the loss of data.

Redundancy   Having one or more backup systems available in case of failure of the main system.

Remote Access   The communications with a computer from a terminal or PC that are separated by a distance. This usually implies a long distance connection over a telephone link.

Repeater   A digital signal regenerator used to extend the length of a physical cable.

Restore   Typically means to put a LAN back into full operation, after a disaster.

RFI   Radio frequency interference.

Router   The hardware and software required to link two subnetworks of the same network together, or to link two remote networks together. The router operates at layer 3 of the OSI reference model.

Spread Spectrum   A radio transmission system that uses multiple frequencies spread across a range of frequencies to transmit information.

Switched 56 kbits/s   A service offering from carriers allowing the customer to dial up communications (primarily data) at speeds of 56,000 bits per second.

T1   A digital transmission technique that multiplexes 24 channels of voice or data (at speeds of 64 kbits/s) across a single four-wire circuit. The base level of the digital hierarchy.

Token   A usage or control packet passed from device to device using a token access method to indicate which device is currently in control of the network or medium.

Token Passing   A collision avoidance technique whereby each station or device passes the control element around the network in a deterministic manner (usually counterclockwise).

Twisted Pair Cable   Two wires of pure copper in a signaling circuit, twisted around each other to minimize the effects of inductance or noise.

UPS   Uninterruptible power supply. An auxiliary power unit for computer or LAN systems that will provide continuous power in the event of a commercial power failure. Typically it is a bank of batteries, but can also consist of a generator (gas, diesel, or natural gas fuel). However, most commercial power failures are of 5 minutes or less.

Warm Site   A facility in between a hot and cold site. Typically some limited systems and capabilities exist.

WAN   Wide area network. A network that links metropolitan, campus, or local area networks across greater distances; usually linked together by common carrier lines.

Index

Alternative sites, 163–167
  hot, 169
  hotels, 167
  options, 164–165
  network, 161
ANI, 88, 100
Asynchronous servers, 73

Backup, 143–159, 196
  floppy disk, 143–145
  optical, 152–154
  procedures, 186
  systems, 143–160
  tape, 146–148, 150–152
Bandwidth, 22–24
  baseband, 23
  broadband, 23–24
Bomb, time, 62
Bomb threats, 190
Bridges, 73–79
Brouters, 73
Browsers, 63
Building inaccessibility, 139–140
Business impact analysis, 36, 188–197

Cable, 23–25
  coaxial, 23–25
  fiber optic, 24–25
  recovery, 116–127
  twisted pair, 24–25
Cable systems, 44–48, 109, 111–132
  backbone, 119
  color codes, 118–119
  cuts, 109, 121–123
  fire damage, 123–126
  isolation, 116
  label, 113, 117
  levels, 45, 47
  patch panel, 120, 122–123
Central office, 74
Communications, 66–103
Communications security, 104–107
Computer manufacturers, 169–170
Connectivity, 66–70
  issues, 66
  LAN–LAN, 66, 68–70

Connectivity (Cont.):
  LAN–WAN, 66, 70
Connectors, 12, 13, 48, 115
  off-keyed, 48
  speciality, 12–13
Controller, cluster, 28, 29
CSMA/CA, 16
CSMA/CD, 16

Data processing, 1–3
  history, 2–3
Departmental recovery, 143
Disaster recovery, 26, 32
  plan, 30, 32
  planning, 26, 30, 33–34
Disk:
  duplexing, 148
  floppy, 146
  mirroring, 148

Electric spikes, 132133
EMI, 109, 117, 130–132
Equipment types, 10
Ethernet, 75, 76, 125

FDDI, 23, 24
Fiber optics, 24–25
Fire, 123–125, 138
Floppy disk, 143–145
Front-end processor, 26–29

Gateways, 73, 81–83, 87

Hackers, 64, 84, 86, 88, 96–100, 102
Hotel, 167–168
Hot site, 169–170

Implementation, 205–206
Inaccessible, building, 139
Interoperabilities, 67
Inventories, 179
  forms, 195

LANs, 1–25
 boundaries, 6–9
 cables, 25, 44–48, 119
 defined, 4–6
 failures, 30–32
 functional parts, 9–10
 media, 24–25
 reasons, 1
 topologies, 14–24
 wireless, 124–126
LAN standards, 178–179
LAN sweep, 153, 156–157
LAN threats, 61–65
LAN to LAN, 66, 68, 103
LAN to WAN, 66, 103
Leased lines, 101–103
Loss, 135–136

Maintenance plan, 222–225
MAN, 23, 71
Management, 34–41
 commitment, 34
 justification, 34–37
 presentation, 38–41
MAU, 21, 109, 110, 131–133
 dropped, 132
 electric spike, 132
 loss, 135–136
 overheating, 133–135
Media, LAN, 24
Media access unit, 13, 110, 121, 131
Mirroring, 148
Modems, 83–103
 communications, 83–85
 dial-back, 97–100
 dial-in, 85–91
 dial-out, 91
 pools, 92–96

Parcel pass, 137
Passwords, 105, 107, 180
Personal computer, 140, 142
Personnel change form, 193
Phases, plan, 207–214
Phone tree, 181, 189
Physical recovery, 108–142
Physical security, 108–109
Plan, 30, 181, 182–185, 196
 action, 196
 developing options, 42–43
 forms, 181, 186–194

Plan (*Cont.*):
 implementation, 198
 maintenance, 222
  control, 224
  distribution, 224–225
  frequency, 223
  modifications, 219–220
 phases, 207–214
 preliminary, 37–38
 presentation, 38–40
 strategies, 33–34
 table of contents, 182–185
 writing, 174–178
Power, 49–54
 systems, 49
 uninterruptible, 51–54
Printers, 55–57, 58
Protect environment, 200
Protection, 44
 surge, 84, 89
Public switched access, 67, 70–72

RAID, 160–161
Recoverable, 163
Recovery:
 cables, 109, 111–131
 communications, 161
 evolution, 3
 network, 161
 physical, 108–109
Restorable, 163
RFI, 109, 130–132
Rodent, 109, 127, 129
Routers, 73, 79–81
R's, four, 210

Salami attacks, 63–64
Satellite communications, 75
Security, 54–57, 104, 108
 communications, 104
 hardware, 54
 physical, 54, 108
 printers, 55–57
 servers, 57
 software, 57
Servers, 11, 14, 110, 136–138
 fire damage, 110, 137–138
 loss, 137
 physical damage, 139–141
 security, 57
 water damage, 138

Shielding, 117–118
Software, 57–61, 143
  security, 57–58
  protection, 59–61
Standards, 178
Strategies, 33, 215–217
  planning, 33
  testing, 215
Surge protection, 84, 89

Tape backup, 146–148, 150–152
Team formulation, 188
  organization, 194
Telephone report, 191
Terminals, 140–142
Testing, 217, 220–221
  disruptive, 220
  frequency, 218
  nondisruptive, 220
  strategy, 215–217
  variation, 221
  who's involved, 217
Threats, 61–65, 190
  bombs, 190
  browsers, 63
  hackers, 64, 86, 88, 97–99
  salami attacks, 63
  tailgaiters, 64, 93
  time bombs, 62
  trap door, 65
  Trojan Horse, 63
  virus, 61
  zappers, 64
Timetable, 177
Token bus, 20–21
Token ring, 19–20
Topologies, 14–22
  bus, 16, 18
  ring, 19–20
  star, 16–17

Training:
  implementing, 205–206
  issues, 200
  materials, 203–205
  plan, 202
  testing, 205
  types, 202
  updates, 206, 219
Twisted pair, 24, 45
  UTP, 24, 45
  STP, 45

UPS, 51–54

Vines, 159
Virtual disk, 155, 157
Virus, 61, 178
Virus prevention, 180
Vortex, 155, 157–159
VSAT, 170

Water damage, 126–127
Wireless, 124–125
Workstations, 110, 140–142
Worm, 152, 156
Writing the plan, 174–201
  backup procedures, 186
  categories, 185
  final checklist, 201
  final review, 199
  master, 175
  subplans, 175
  table of contents,
    182–185
  timing, 177

Zappers, 64

ABOUT THE AUTHOR

Regis J. (Bud) Bates, Jr. has more than 27 years of experience in telecommunications and management information systems, with specific expertise in end-user management, system integration, disaster recovery and avoidance, strategic planning, and cost reduction and containment. Mr. Bates is the president of TC International Consulting, Inc. of Phoenix, Arizona. The firm specializes in voice and data communications, technical training development, and instruction. He is an active lecturer and the author of numerous trade articles and books, including *Disaster Recovery Planning: Networks, Telecommunications, and Data Communications* (McGraw-Hill, 1991); *Wireless Networking Communications: Concepts, Technology, and Implementation* (McGraw-Hill, 1993); and *Introduction to T1/T3 Networking* (Artech House, 1992).